Mickael Mehou-Loko

Qatar's soft power in Europe and the Arab world

Mickael Mehou-Loko

Qatar's soft power in Europe and the Arab world

ScienciaScripts

Imprint
Any brand names and product names mentioned in this book are subject to trademark, brand or patent protection and are trademarks or registered trademarks of their respective holders. The use of brand names, product names, common names, trade names, product descriptions etc. even without a particular marking in this work is in no way to be construed to mean that such names may be regarded as unrestricted in respect of trademark and brand protection legislation and could thus be used by anyone.

Cover image: www.ingimage.com

This book is a translation from the original published under ISBN 978-3-8417-2635-3.

Publisher:
Sciencia Scripts
is a trademark of
International Book Market Service Ltd., member of OmniScriptum Publishing Group
17 Meldrum Street, Beau Bassin 71504, Mauritius
Printed at: see last page
ISBN: 978-620-3-14555-7

Copyright © Mickael Mehou-Loko
Copyright © 2020 International Book Market Service Ltd., member of OmniScriptum Publishing Group

Acknowledgements

I would like to thank Mr. Verschuuren who guided me throughout this brief.

I would also like to thank Mr Bitar for his sound advice.

Finally, I would like to thank Mr Ennasri for granting me an interview.

Table of abbreviations

CNT: National Transitional Council CNS: Syrian National Coalition

NGO: Non-Governmental Organisation UN: United Nations Organisation

OPEC: Organisation of the Petroleum Exporting Countries NATO: North

Atlantic Treaty Organisation GDP: Gross Domestic Product

PSG : Paris Saint Germain

QIA: Qatar Investment Authority

Introduction

"When I travelled in Europe when I was young, in the airports, the police were always asking me:

but where is Qatar? ", Sheikh Hamd Bin Khalifa Al-Thani

Qatar is a country in the Middle East with an area of 11,586 km² and a population of 1,758,800. The capital of Qatar is Doha. The majority religion is Sunni Islam, practised by 95% of the population, and the official language is Arabic. Qatar is a constitutional monarchy ruled by Hamad Khalifa al - Thani, Emir of Qatar and head of the Qatari state between 1995 and 2013. His son, Tamim succeeded him on 25 June 2013. The country is divided into 7 municipalities. The countries bordering Qatar are the island of Bahrain to the north-west and Saudi Arabia to the south. To the north, Qatar has access to the Persian Gulf.

Oil exploitation began in 1979. This was the birth of a singular economic model of a constitutional monarchy with a rentier economy based on the export of hydrocarbons. There is a large immigrant population and Qatar is almost in a situation of full employment. Qatar's economic growth has been very rapid in 20 years. However, this economic growth has its limits.

Qatar is not yet a developed country. Its economic development model has limits. First of all, Qatar's economy is dependent on foreign countries. Its population is composed of 80% immigrants in 2010. Managers come from abroad and the Qatari population is not well trained.

The Middle East is the region of all cleavages. Civilisational cleavages first of all with an opposition between East and West on themes such as human rights, women's rights, democracy and freedom of expression. The cleavages manifest themselves on different scales. For example, on a global scale, the opposition between the Catholic and Muslim faiths is a cleavage that dates back to the Holy Wars of the Middle Ages. In the Near and Middle East region, the geopolitical and religious opposition between Israel and the Muslim countries of the Middle East and Maghreb was previously symbolised by the Arab-Israeli conflicts and now by the Israeli-Palestinian conflict. Among Muslim countries, there are rivalries between Sunni and Shiite countries. Iran is the leader of the countries of the Shia religion, and its foreign policy has a great influence on the Shia minorities in other Middle Eastern countries, notably in Lebanon with Hezbollah and in Syria. Iran and Israel clashed in negotiations between Iran and the United States over Iran's uranium enrichment programme, which resulted in a historic agreement. Finally, the Sunni Muslim world is looking for a geopolitical leader. Until now, Turkey, Egypt, Iraq and Saudi Arabia have taken turns in this geopolitical leadership role. Qatar would like to assume this role in the future.

Qatar is surrounded by two regional powers. It shares many characteristics with Saudi Arabia, a Wahhabi Sunni Muslim and pro-American country on a par with Qatar. On the other hand, Iran is a Shiite Muslim country and hostile to the Americans even if economic and diplomatic relations between the two countries are tending to normalise. Since its independence in 1971, Qatar has felt threatened by these two countries. The border disputes with Saudi Arabia and Bahrain were only resolved in 2001. While Qatar has 11,800 men and spends 1.99% of its GDP on defence, Saudi Arabia has 124,000 soldiers and its defence budget represents 8.26% of its GDP. As for Iran, its military strength is 523,000 men and its defence budget represents 2.40% of its GDP. Thus, Qatar is far from competing with its neighbours as a military power. The example of Iraq's invasion of Kuwait in 1990 makes it fear for its security. In order to make its territory safe, Qatar is counting on the protection of the United States. In 2002, the United States set up a military base in El-Udeid. Qatar became the command centre for

American military operations in the Middle East.

Now aware of its limits, Qatar no longer wants to be dependent on hydrocarbon resources that will diminish in the coming decades. The Arab revolutions have shown the limits of countries such as Tunisia or Libya whose economy is based on rent. Moreover, without military power equivalent to Iran or Saudi Arabia, soft power through economy and sport seems to be the only way for Qatar to compete with its two neighbours.

Qatar is defining a new development strategy in 2010. In Qatar Vision 2030[1], one of the pillars of this new strategy is direct investment abroad. It is a question of investing the oil manna in large companies and multinationals in the world. In this way, Qatar seeks to reap benefits from its investments and extends its economic influence in Europe where its investments are best placed and most likely to be profitable.

Qatar is implementing a policy of multidimensional influence. Qatar has an admiration for Europe and particularly for France, whose model of economic and cultural development is the goal Qatar would like to achieve by 2030. Investments enable Qatar to establish strong relations with major powers and to influence international negotiations. Qatar invests in France thanks to its sovereign investment fund, the Qatar Investment Authority (QIA). But Qatar does not only invest in France. It invests in other European countries such as the United Kingdom, France and the United States.
-United Kingdom and Germany.

How can a country that is organising the World Cup in 2022 and investing in the West at the same time finance jihadist Islamist groups in Mali and Syria?

Qatar is the subject of passionate debate. The objective of this study is to

---

[1] http://www.gsdp.gov.qa/portal/page/portal/gsdp_en/qatar_national_vision/qnv_2030_document/QN V2030_English_v2.pdf, accessed 25 March 2013

analyse in a rigorous and objective manner the origins and logic of Qatari investments in Europe and more precisely in France. In a second part, an analysis of Qatar's foreign policy in the Middle East, the Maghreb and Mali since 1995 is necessary to show how Qatar has lost its image of modernity and neutrality.

Part 1: Economic diversification as a source of investment in Europe?

This part is an analysis of Qatari investment strategy in Europe. The diplomatic ties that have been created between Qatari and French leaders have facilitated the opening of French markets, especially during the mandate of Nicola Sarkozy between 2007 and 2012. The economic crisis began in 2008 and led to an acceleration of Qatari investments in France in three sectors in particular and with different strategies: real estate, industry and sport.

A- A rent economy based on the sale of hydrocarbons

Over the last twenty years, Qatar's emergence on the economic scene has been dazzling. Between 1993 and 2003 GDP per capita doubled while GDP tripled over the same period[2]. Between 2003 and 2011, GDP per capita increased by 800%, placing Qatar at the top of the 2011 ranking of GDP per capita of the States[3]. Throughout the first decade of the 21st century, Qatar recorded an average economic growth of 21.8% per year[4]exploitation of hydrocarbon reserves shortly after the country's independence in the 1970s.

---

[2] Nabil Ennasri, L'énigme du Qatar, Iris Editions 2013, p. 67.
[3] Ibidem
[4] Pascal Boniface, The Strategic Year 2013, Analysis of international challenges

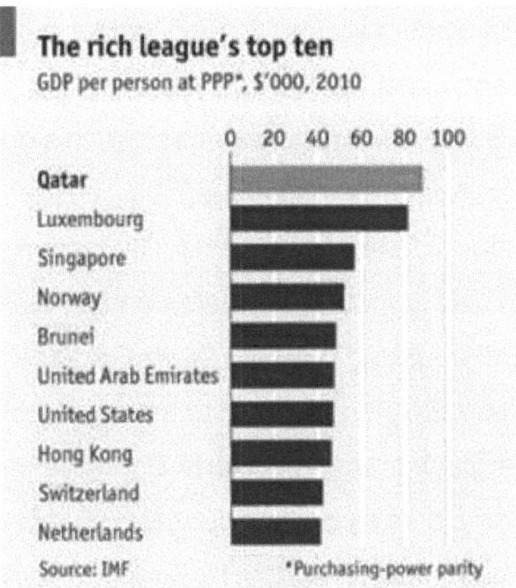

http://www.economist.com/node/21536659 Figure 1: Top 10 GDP per capita

The exploitation of hydrocarbons, at the origin of Qatari economic power

Oil does not give Qatar a comparative advantage on the world market. Indeed, Qatar is the smallest oil producer in the Organisation of Petroleum Exporting Countries (OPEC) with reserves representing 2.3% of the world's oil reserves [5]. The North-Dome, the world's largest offshore natural gas reserve, was discovered in 1971. It is a deposit of 6,000 square kilometres [6]divided between Qatar and Iran with an exploitation period of more than a century. In a region where oil is abundant, Qatar is the only country in the Gulf to have natural gas reserves on its territory.

Qatar Gas was founded in 1984 with the aim of making a profit from the exploitation of natural gas. Qatar Gas is still today the source of financing for the Qatar National Vision 2030 project. Through Qatar Gas, the Qatari

---

[5] CIA World Factbook 2012
[6] Nabil Ennasri, L'énigme du Qatar, Iris 2013, p/ 61

authorities work with major multinationals in the oil and gas sector such as Shell, Exxon and Total. In 1936, Qatar signed a partnership with Total as part of the British protectorate. Qatar seized this opportunity to get out of its dependence on Anglo-Saxon capital.

Until 1990, Qatar did not have the financial resources to exploit natural gas. An agreement was signed with Total to enable Qatar to no longer be solely an oil exporting country. In 1994, Qatar Gas opened its capital to foreign capital. Total and Exxon contribute to the launch of the Qatar Gas project.

1.   Natural gas extraction in North Dome began in 1997. It is managed by Exxon. North Dome's natural gas reserves represent 14% of[7]the world's conventional gas. The Qatargas II, III and IV projects follow the logic of partnership diversification. Matsui and Marubeni become partners, from Qatar for the extraction of natural gas. Moreover, Qatar is at the origin of a regional project in agreement with its neighbours. The Dolphin project aims to bring Qatari gas to the United Arab Emirates and Oman via a pipeline. The Qatari authorities are investing 70 billion dollars[8] in the Ras Laffan site so that it can become the world's first industrial-gas complex.

In 2010 Qatar becomes the leading exporter of liquefied gas with 77 million tonnes of production[9]. Qatar has been able to diversify its exports by meeting the growing demand from emerging countries. Demand in China for liquefied natural gas increased by a third in 2011[10]. As for India, its imports of liquefied natural gas are expected to triple[11] . Qatar will have to compete with Australia which has invested 170 billion dollars[12] in this field since 2007. Hydrocarbons represent 95% of Qatar's exports, 75% of its budget revenue, 32% of its GDP and 28% of its national wealth[13].

---

[7] Nabil Ennasri, L'énigme du Qatar, Iris 2013, p. 62 and the International Energy Agency
[8] Nabil Ennasri, L'énigme du Qatar, Iris Editions 2013, p. 62.
[9] Ibidem, p. 63

[10] Ibidem p. 64
[11] Ibidem p. 64
[12] Ibidem p. 65
[13] Ibidem p. 65

The entire Qatari economy is thus based on the export of hydrocarbons. Even if the profits generated by the export of hydrocarbons are substantial, Qatar cannot develop without diversifying its activities, all the more so as the prospect of Peak Oil calls into question the Qatari development model. This is what is at stake in the Qatar National Vision for 2030. The financial manna coming from hydrocarbons enables Qatar to create sovereign funds. What are its vectors of Qatari economic power? How are they structured and organised? Are Qatari sovereign funds endowed with a significant financial envelope on a global scale?

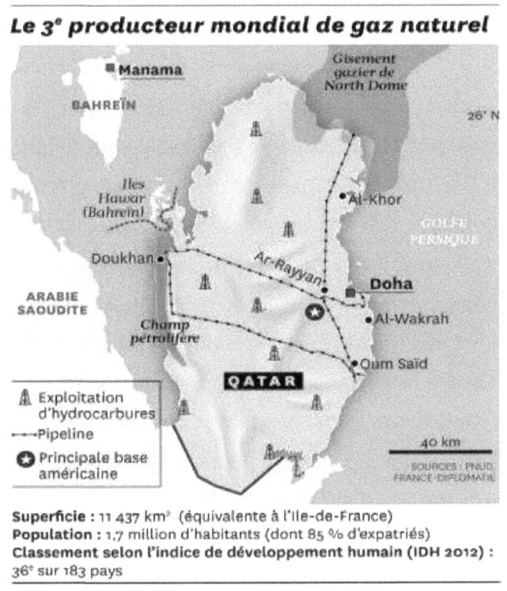

Figure 2: Map of the North Dome

http://www.economist.com/node/21536659

The Qatar Investment Authority, a vector of Qatar's economic strength

The Qatar Investment Authority is the Qatari sovereign wealth fund. It is responsible for developing investments abroad with the aim of making the country's financial surpluses profitable. The Qatar Investment Authority "develops, invests and manages the State's reserve funds in accordance with the policies, plans and programmes approved by the Supreme Council"[14]. This sovereign fund is structured by subsidiaries, each dealing with a sector.

Qatar Holding was created in 2006. Its mission is to invest in large companies and multinationals. This investment fund is headed by Prime Minister Ben Jassim Al-Thani. The Qatar Holding has assets of around 65 billion dollars[15]. The real estate sector is entrusted to the Qatar Diar Real Estate company. Investments in sport abroad are entrusted to Qatar Sport Investment, created in 2010. Sheikh Tamim Al-Thani is the sole decision-maker for Qatar's investments in sport. Finally, the luxury sector is entrusted to the Qatar Luxury group under the leadership of Sheickha Mozah. The Qatar Luxury group is independent of the Qatar Investment Authority but is part of the Qatar Foundation headed by Sheickha Mozah. The Qatari Sovereign Wealth Fund is structured and organised in such a way that decision-making power is quick and efficient.

However, the economic power of Qatar should be put into perspective. Indeed, the assets of the Qatar Investment Authority are estimated at 100 billion dollars[16]Qatari sovereign fund is ranked twelfth in the ranking of sovereign funds, far behind the sovereign fund of Abu Dhabi whose assets exceed 800 billion dollars[17]. The Qatari investment strategy is more recent

---

[14] Article 5 of the Princely Decree number 22 of 2005

[15] http://www.lefigaro.fr/societes/2010/05/09/04015-20100509ARTFIG00224-le-qatar-achete-harrods-a-mohammed-al-fayed.php consulted on April 27th

[16] Nabil Ennasri, L'énigme du Qatar, Iris Editions 2013, p. 68.

[17] Ibidem, p. 69

than that of its Gulf neighbours which already occupies a major part of the market in the United States. The Qatari economic strategy makes Europe the field of predilection for its investments.

Ranking of the top 25 sovereign wealth funds in the world in 2012		
State	Sovereign wealth funds	Assets (bn USD)
WATER	Abu Dhabi Investment Authority	627
Norway	Government Pension Fund Global	611
China	SAFE Investment Company	568
Saudi Arabia	SAMA Foreign Holdings	533
China	China Investment Corporation	440
Kuwait	Kuwait Investment Authority	296
China - Hong Kong	Hong Kong Monetary Authority Investment Portfolio	293
Singapore	Government of Singapore Investment Corporation	248
Singapore	Temasek Holdings	157
Russia	National Welfare Fund	150
China	National Social Security Fund	135
Qatar	Qatar Investment Authority	100
Australia	Australian Future Fund	80
WATER	Investment Corporation of Dubai	70
Libya	Libyan Investment Authority	65

Kazakhstan	Kazakhstan National Fund	58
WATER -Abu Dhabi	International Petroleum Investment Company	58
Algeria	Revenue Regulation Fund	57
WATER - Abu Dhabi	Mubadala Development Company	48
South Korea	Korea Investment Corporation	43

Figure 3: Ranking of the top 25 sovereign wealth funds in 2012

http://www.fondssouverains.com/article-classement-des-fonds-souverains-par- actifs-geres-106519471.html

Qatar's economic strategy is inspired by that of Kuwait in the 1980s. Kuwait pioneered a strategy of investing the oil windfall abroad in order to obtain income from investments abroad that exceeds the revenue from the sale of oil. In the Qatar National Vision 2030, outward investment is one of the pillars that would enable Qatar to become a developed country by 2030. Qatar is implementing an effective economic strategy through its sovereign wealth funds.

Le Qatar a poussé ses pions en Europe

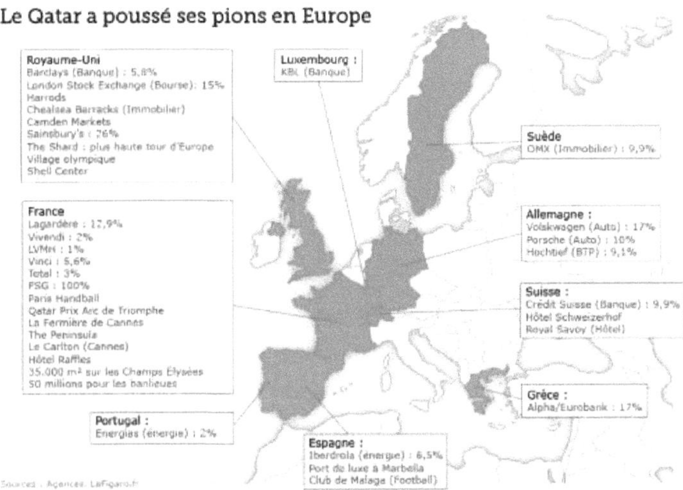

Figure 4: Qatar's investments in Europe

http://www.lefigaro.fr/conjoncture/2012/06/22/20002-20120622DIMFIG00526-la- octopus-power-qatar-in-the-world.php

Why is Qatar investing in Europe?

Qatar's investment strategy has had to adapt to the global market. Saudi, Kuwaiti and Emirati investments being massive in the United States, Qatar has turned towards Europe. Qatar's investments in Europe are a means of ensuring its security against its militarily more powerful neighbours by increasing its soft power[18]. The agreements signed between Qatar and the United States are of a military nature, such as the installation of the largest American base in Qatar. In Europe, Qatar's strategy is to invest in the best sectors of each country to ensure profits and to be able to resell once the economic crisis is over.

In the United Kingdom, Qatar's investment strategy focuses on real estate in particular. It is in the United Kingdom that Qatari investments are the most important while it is in France that these investments are the most mediatised. However, they remain targeted on a few sectors such as real estate and finance. Thus, Qatar has owned Harrods shops since 2010. Qatar Holding bought Harrods from Mohammed al-Fayed for the sum of 1.7 billion euros[19]. Harrods has the triple advantage of being a luxury shop, a bank and a real estate agency. It welcomes 15 million people every year in its 90,000 m² London department stores'. Also in London, Qatar acquired the Olympic Village in 2011 for 663 million euros[20] through its subsidiary Qatar Diar. In partnership with the British real estate group Delancey, the construction of more than 2,000 houses is planned[21] Shard, built in London and completed in June 2012. The tower is 310 metres high and cost around 1.8 billion euros, 80% of which was financed by four Qatari banks and the

---

[18] Soft power: dissemination of an attractive economic and political model, Nye's "cultural attraction" based on the mastery of telematic channels and in particular the media contribute to the capacity of attraction.

[19] http://www.lefigaro.fr/societes/2010/05/09/04015-20100509ARTFIG00224-le-qatar-achete-harrods-a-

mohammed-al-fayed.php consulted on 10 May 2013

[20] http://www.europe1.fr/Economie/JO-2012-le-village-olympique-vendu-au-Qatar-666313/ consulted on 15 May 2013

[21] ibidem

Qatar Diar[22] investment fund. It will be a tower with multiple functions: offices, hotels, flats and restaurant. In these last two acquisitions, it is noticeable that Qatar is not only buying existing buildings. The construction projects completed or to come are proof of a real long-term investment strategy of Qatar in the United Kingdom. Moreover, these acquisitions are taking place in a context of economic crisis and thus create activity in the construction sector. Since 2007, Qatar has held a 26% stake in the shares of Sainsbury's supermarkets, the UK's third largest retail group. In the finance sector, Qatar owns 20% of the shares of the London Stock Exchange and 7% of Barclays Bank. These substantial investments are very well received in England. The economic crisis has enabled Qatar to increase its presence in the United Kingdom.

In Germany, Qatari investments target the industrial sector. In 2010, Hochtieff, a construction group, sold 9.1% of its shares to Qatar Holding. In the automobile industry, Qatar holds 10% of Porsche and 17% of Volkswagen. 7 billion of investments were announced as soon as Volkswagen was bought[23] out.

In Spain, Qatar has owned 6.16% of Iberdrola since March 2011. In February 2012, Qatar's share reached 8.4%, representing an investment of more than 2 billion dollars in this Spanish electricity supplier. Qatar has benefited from Iberdrola's 16% drop on the stock market, reaching its lowest level since 2008. Qatar owns 2% of Energias' shares in Portugal in the energy sector as well.

Qatar invests in Europe in a well thought-out way following a "buy and hold" strategy. Buy and hold" consists in stabilising its investments by accepting "lock in" clauses so as not to resell their shares for a minimum of one or two years. The objective is to sit on the Board of Directors for several years. [24]

---

[22] http://www.lemonde.fr/proche-orient/article/2012/02/25/le-qatar-s-offre-la-plus-haute-tour-d-europe_1648127_3218.html consulted on 15 May 2013

[23] http://www.20minutes.fr/article/557135/Economie-Le-Qatar-engagera-plus-de-7-milliards-dans-Porsche-et-Volkswagen.php, accessed 16 May 2013

[24] C.Chesnot,G.Malbrunot: Qatar les secrets du coffre-fort, Michel Laffon 2013 p.258.

How did France become a privileged partner of Qatar in Europe?

Investment is facilitated by long-standing diplomatic and personal ties between Qatar and French leaders that were strengthened in France during Nicolas Sarkozy's presidency and also following the economic crisis of 2008.

Relations between France and Qatar are long-standing. They are the fruit of a common will to emancipate themselves from the tutelage of their powerful allies or neighbours: the United States for France and Saudi Arabia for Qatar. France's desire to have a foreign policy independent of the United States is expressed through Gallicism-mitterrandism characterised by a singular Arab policy. The foundations of this singular Arab policy are the criticisms made of Israel after the Six Day War of 1967 by President De Gaulle. France defined itself as a "great middle power" capable of dealing on an equal footing with the two Cold War superpowers, the United States and the Soviet Union. This expression coined by Valéry Giscard D'Estaing in the 1970s means that during the Cold War, France was not a superpower like the USSR and the United States, but that it remained superior to the other middle powers of the 1970s: Japan and Chile, for example. It has had atomic weapons since the mid-1960s. It sits on the UN Security Council on a permanent basis and has a right of veto. Moreover, France is the second largest diplomatic network in the world after the United Kingdom, which gives it considerable influence[25]. Following the example of France, which has sought to emancipate itself from American tutelage, Qatar is seeking to emancipate itself from Saudi tutelage. How did Qatar become a "great middle power" in the Middle East?

Qatar's desire to have a foreign policy independent of Saudi Arabia has been expressed since 1995. Qatar initially became the ally of Iran and Syria and maintained privileged links with Israel. Its two economic and strategic partners therefore share a desire to free themselves from the tutelage of their powerful allies.

---

[25] ibidem

In the 1990s, relations between Qatar and France were of a strategic and military nature. Iraq, the leader of the Middle East, weakened after its war against Iran from 1980 to 1988 and the Gulf War in 1990 and 1991. The Middle East is then in a period of transition. France must diversify its strategic alliances in the Middle East. Defence agreements were signed between Qatar and France in 1994 and 1998. Thanks to these agreements, France became the leading supplier to the Qatari army (80% of its military equipment)[26]. The value of the equipment order under the 1998 agreements reached 80 million dollars[27]. This includes armoured vehicles from GIAT Industries, Mirage 2000-5 and Milan, Exocet and MICA missiles. In addition, military exercises are regularly scheduled. Qatar accepts the presence of French troops on its territory to ensure the security of its gas platforms located a few kilometres from Iran.

From a diplomatic point of view, relations between the two countries have intensified since the presidency of Chirac, who visited Qatar nine times during his two terms of office between 1995 and 2007 (one seven-year and one five-year term). However, the privileged strategic partner is still Saudi Arabia. On the Qatar side, the main strategic ally is still the United Kingdom at the beginning of the 21st century. Symbolic gestures increase the degree of cooperation between Qatar and France. The Emir of Qatar sent his son Joann to Saint Cyr in 2003 while his elder sons had gone to the Royal Military Academy of Sandhurst in the United Kingdom. From the presidency of Sarkozy between 2007 and 2012, diplomatic relations between Qatar and France increased until they reached the "all-Qatar"[28] level. The Emir of Qatar was received at the Elysée Palace three weeks after Sarkozy's election. He is the first Arab head of state to be received by the French President, a sign heralding close strategic cooperation between the two countries. A contract for the purchase of 80 Airbus A-350s is concluded for

---

[26] Nabil Ennasri, L'énigme du Qatar, Iris 2013, p 151

[27] ibidem

[28] Nabil Ennasri, L'énigme du Qatar, Iris 2013 editions. "All Qatar": expression used by Mr. Ennasri to describe the almost exclusive relationship that France had with Qatar during the Presidency of Nicolas Sarkozy (2007-2012).

16 billion euros[29]. On 14 July 2007, the Emir of Qatar was invited to the military parade. A few months after the liberation of the Bulgarian nurses, Qatar paid 460 million dollars to[30] the families of the victims. This cooperation allows France to take advantage of the good relations between Qatar, Syria and Hamas to weaken Iran. Bashar El Assad is reintroduced on the international scene and pays an official visit to France in July 2008 and again in December 2010. Syria is an essential element for the credibility of the Union for the Mediterranean. Qatar plays a leading role in the organisation and financing of the UfM. Following Morocco's refusal to participate in the Union for the Mediterranean, Qatar has become the Arab guarantee that France needs for the Union for the Mediterranean to appear credible.

Another parameter to be taken into account is the French-speaking world of the Emir of Qatar. Emir Hamad Al Thani learned French during his years of military service in the 1980s thanks to Professor Larrieu. This teaching was then passed on to Sheikha Mozah, the only wife with whom the Emir appeared in public, and to his sons. Qatar is showing signs of its love of culture with the inauguration on 15 January 2008 of the Franco-Qatari Voltaire High School in Doha in the presence of President Sarkozy [31]. The French Secular Mission is in charge of the pedagogical mandate and the programmes are approved by the French National Education. Qatar joined La Francophonie as an associate member in October 2012. In exchange, Qatar has made French one of the official languages of its school curriculum.

---

[29] Nabil Ennasri, L'énigme du Qatar, Iris 2013, p.155

[30] Ibidem, p.153

[31] Coustillière Jean-François, "Qatar: chance or threat to French interests? "Confluences Méditerranée, 2013/1 No. 84, pp. 87-100. DOI: 10.3917/come.084.008

C- Qatar's investments in France

Qatar's investments in France are not the fruit of the whims of a Francophile emir. They obey a simple logic: profitability. Thus, Qatari sovereign funds invest in everything that makes the excellence of French soft power: luxury real estate, industry, sport and the media. Like Kuwait since the 1980s, Qatar benefits from tax advantages for investing in France. In 2008, Qatar and France signed an amendment to the tax treaty of 12 January 1993 aimed at improving "the attractiveness of France for Qatari investors, particularly in the real estate sector"[32]. The text was ratified in February 2009 by Parliament. It grants Qatari investors an exemption from taxes on real estate capital gains and capital gains realised in France. In addition, Qatari nationals are exempt from wealth tax "on property located outside France for a period of five years" after becoming resident in France. These tax advantages open up the real estate, industrial and sports markets in Qatar. Qatar's prestigious investment strategy has three long-term objectives. Investments in real estate serve to obtain capital gains on resale when the crisis is over. Investments in French industry serve to attract know-how with the aim of developing Qatar by 2030, when Qatar wants to become a developed country. Finally, investments in French sport show that Qatar wants to strengthen its soft power and become a credible organiser of the 2022 World Cup.

---

[32] http://www.impots.gouv.fr/portal/deploiement/p1/fichedescriptive_2100/fichedescriptive_2100.pdf consulted on 2 June

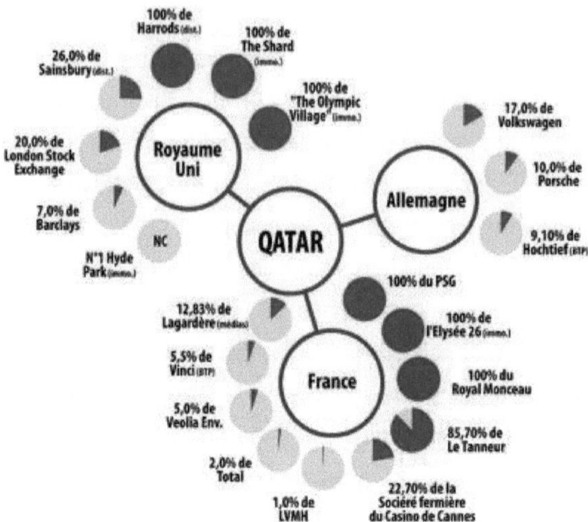

Figure 5: Diagram of Qatar's investments in France, the United Kingdom and Germany

http://www.latribune.fr/entreprises-finance/banques-finance/industrie-financiere/20120326trib000690322/le-qatar-ne-rachete-pas-que-la-france.html

Investments in luxury real estate in France

These are investments by the Qatari sovereign fund Qatar Diar Real Estate Company and its subsidiaries Barwa Real Estate and Katara Hospitality. The real estate market has four advantages. The first is the increase in demand and the attractiveness of the sector: more and more billionaires from emerging countries are attracted by this market. The second advantage is the excellence of Paris in the luxury sector. The value of luxury and heritage properties is unlikely to fall. The third advantage is the possibility of making considerable capital gains once the economic crisis is over. Finally, the fourth advantage is the possibility of acquiring spaces frequented by important politicians and business leaders for lobbying purposes.

Elypont SA is in charge of the management of the Qatari portfolio and the orientation of Qatari investments on French territory. A former Lebanese diplomat manages Elypont and he hopes to obtain "a return of 4% to 5%"[33] on his investments, which will only target investments of more than 100 million euros[34].

In 2007, the Royal Monceau Hotel was bought for 250 million euros[35] following a partnership between Barwa Real Estate and Alexandre Allart[36] . It is one of the meeting places of the Parisian elite. In 2008 Katar Hospitality bought the former international conference centre on Avenue Kleber. The building becomes a luxury hotel under the Asian Peninsula brand. The cost of the renovations is estimated at 210 million euros for the rooms[37] . In 2012, Qatar Diar wins the tender to buy the Virgin building from Groupama, which was badly affected by the Greek crisis. The amount of the transaction is more than €500 million. This is the third largest transaction in France in the

---

[33] Nabil Ennasri, L'énigme du Qatar, Iris 2013, p.160

[34] ibidem

[35] Ibidem, p.161

[36] http://www.lefigaro.fr/societes/2008/10/01/04015-20081001ARTFIG00367-le-qatar-s-invite-au-majestic-a-cannes-.php accessed 3 June 2013

[37] Nabil Ennasri, L'énigme du Qatar, Iris 2013, p.161

last ten years. In June 2012, the Katara Hospitality will become, according to

certain sources[38] the purchaser of four luxury hotels that were owned by private equity firm Starwood Capital. The brother of Prime Minister Hamad Ben Jassim is the head of Katara Hospitality. They are the Louvre Hotel, the Concorde La Fayette Hotel in Paris, the Martinez Hotel in Cannes and the Palais de la Méditerranée in Nice. Indeed, Qatar does not only invest in Parisian real estate. On the Côte d'Azur, it has owned 27% of the shares of the company that has been running the municipal casino in Cannes since 2008[39]. This company manages the two casinos Barrière Croisette and Les Princes, the Majestic hotel and the Gray Albion. Finally, in 2012, Qatar will become the purchaser of the Neo building for 300 million euros[40], which houses the Figaro and several services of the US embassy.

It is also important to underline the private acquisitions made in France by members of the royal family and their relatives. The Emir of Qatar Hamad Al-Thani owns a luxurious duplex in Place Rivoli. He owns a residence in Marnes La Coquette in western Paris and a villa in Château Neuf Grasse in the Alpes Maritime. He acquired the Hotel d'Evreux and two nearby hotels for 250 million euros[41] . Prime Minister Ben Jassim bought the Kinsky hotel for 28 million euros [42]. In 2008, the Emir's younger brother bought the Hotel Lambert on the island of Saint Louis for 80 million euros, [43]a former property of the Baron de Rothschild. In 2012, a relative of the royal family Ghanem Ben Saad Al Saad buys seven Intercontinental hotels for 450 million euros[44]. Among them is the Carlton, a prestigious hotel on the Cannes

---

[38] http://www.lefigaro.fr/immobilier/2012/06/22/05002-20120622ARTFIG00525-le-qatar-s-offre-quatre-hotels-de-luxe-en-france.php, accessed 3 June 2013

[39] Nabil Ennasri, L'énigme du Qatar, Iris 2013, p.161

[40] http://www.lemonde.fr/proche-orient/article/2012/06/23/le-qatar-rachete-l-immeuble-du-figaro_1723706_3218.html, accessed 4 June 2013

[41] Nabil Ennasri, L'énigme du Qatar, Iris 2013, p.161

[42] Ibidem, p.162

[43] Nabil Ennasri, L'énigme du Qatar, Iris Editions 2013, p. 162.

[44] Ibidem p. 163

Croisette. In 2013, the Printemps is bought by the Emir of Qatar for 1.6 billion euros[45]. It is a gift for his second wife Sheikha Moza. The objective is to double the turnover of Le Printemps and reach 1.4 billion euros in 2019[46]. On

In addition, three shops will open, including one at the Carrousel du Louvre in 2014. Qatar plans to invest 270 million over five years[47].

Investments in industry in France

France has three undeniable assets in the industrial sector. Firstly, France is renowned for having the most successful companies in the world. Second, France has a healthy environment conducive to investment. Finally, according to the UNCTAD ranking, France is one of the best host countries for foreign funds. According to Ernst & Young, France ranks second in Europe just after Germany for the first half of 2012 .48

The Qatari investment fund Qatar Holding is in charge of investing Qatari funds in the best French companies. In 2009 Qatar Holding became the second largest shareholder of the Vinci group behind the employees by buying Cegelec for €1.2 billion[49]. Qatar thus owns 5.5% of the [50]shares of Vinci, the world leader in construction and public works. This is one of the largest investments made in a CAC40 company. In exchange, Vinci has been selected to build the 5 billion euro bridge-digital project between Bahrain and Qatar[51]. In addition, the Vinci group will build some of the nine stadiums planned for the 2022 football World Cup. In 2010, Qatar will buy

---

[45] Kira Mitrofanoff, " Petites combinaisons et grandes ambitions au Printemps ", Challenges, number 350 from 20 to 28 June 2013, pp. 58-59

[46] Ibidem

[47] Ibidem

[48] http://www.challenges.fr/economie/20121113.CHA3007/classement-la-france-2eme-des-investissements-directs-etrangers-en-europe.html, accessed 10 June 2013

[49] Nabil Ennasri, L'énigme du Qatar, Iris Editions 2013, p. 165

[50] ibidem

[51] ibidem

5% of Veolia's shares for €650 million[52]. It thus becomes the group's fifth largest shareholder. This entry into Veolia's capital includes an industrial partnership. Veolia will participate in the implementation of water and waste management infrastructures for the emirate's new developing cities. In March 2012, Qatar Holding will hold 1.03% of the shares of LVMH, the world leader in luxury[53] goods. A few weeks later, Qatar Holding's share in the Lagardère group will increase from 7.6% to 10%[54] . Qatar Holding is the leading shareholder of the group which owns the radio station Europe 1 and the newspaper Figaro. However, Qatar was turned down when it tried to take a stake in EADS. Through Lagardère, which holds 7% of EADS' capital, Qatar Holding indirectly enters EADS' capital[55]. In April 2012, Qatar Holding will invest 2 billion to increase its stake in the Total group to 3%[56]. Qatar is Total's second-largest producer and its main source of production.

Qatar's investments are not only aimed at large companies. A €300 million aid package is planned to help small and medium-sized enterprises (SMEs). This sum will be financed by Qatar Holding and the Caisse des Dépôts in order to "invest in French SMEs, in business sectors with strong growth potential that are of interest to both Qatar Holding LLC and the Caisse des Dépôts[57]".

Investment in sport and media in France: case study: football in France

Qatar's investments in sport in France emanate from the will of Crown Prince Sheikh Tamim Al-Thani. He is the president of Qatar Sport Investment created in 2011. Qatar Sport Investment's investments in France are part of a more global logic. Indeed, Qatar has been designated

---

[52] ibidem

[53] Ibidem, p. 164

[54] Ibidem p. 165

[55] Nabil Ennasri, L'énigme du Qatar, Iris Editions 2013, p. 165

[56] ibid., p. 166

[57] http://www.usinenouvelle.com/article/la-cdc-et-qatar-holding-vont-investir-300-millions-d-euros-dans-des-pme.N187117 consulted from 8 August 2013

as the organizer of the 2022 FIFA World Cup and the 2015 Handball World Cup. The World Cup will represent an investment of 200 billion euros as shown in figure 6. This World Cup will be the most expensive in history because everything remains to be built in Qatar in terms of infrastructure. Qatar also plans to organise the Olympic Games in the 2020 decade. However, although Qatar has the financial means to organise these high-level sports events, it lacks legitimacy with a demanding Western and popular public. Qatar must make itself known in the world of sport. Once again,

Qatari leaders invest according to profitability criteria and avoiding reckless spending.

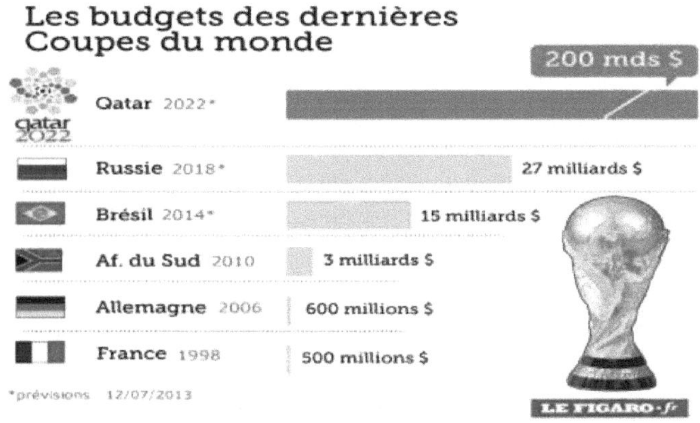

Figure 6: Projected costs of organising the 2022 World Cup
http://www.lefigaro.fr/sport-business/2013/07/12/20006-
20130712ARTFIG00355- football-why-qatar-exchange-again-200-billion-
dollar.php

The French football market seemed to be the most affordable for Qatar. Indeed of the five championships recognised as the best in Europe (England, Spain, Italy, Germany and France) France is a competitive market and has potential to develop. The value of clubs and TV rights is relatively low compared to their European neighbours. This comparative advantage is due to a lack of sporting performance at continental level of French clubs and consequently to its lack of attractiveness. Qatar Sport Investment is entering a promising market that has been little exploited. It is inspired by another Italian owner and politician, Silvio Berlusconi who bought the Milan club before acquiring the TV rights in Italy. Three stages stand out in Qatar Sport Investment's strategy to establish itself in French football: the purchase of the club in the French capital, the acquisition of the television rights of football matches through Bein Sport and finally the development of sponsorship through the Burda brand.

The first step for Qatari investments in France is to look for a club with potential that is conducive to making profits. Qatar Sport Investment surrounds itself with competent advisors such as the son of Michel Platini, the legendary French football player of the 1980s. Paris seems to be a logical choice in view of its assets and attractiveness. As the world's leading tourist destination, it is considered the capital of luxury and an important place for culture and the arts. After a first failure in 2006 to the benefit of an American investor, Colony Capital, which obtained 70% of the Parisian club, Qatar retains its chance in 2010 while the Parisian club is indebted to the tune of 50 million[58] . Negotiations take place in the presence of the highest levels of the State. Indeed, President Sarkozy, also a supporter of Paris Saint Germain, organised a meeting at the Elysée Palace between Cheick Tamim and Sébastien Bazin, President of Colony Capital, the majority shareholder of Paris Saint Germain. Moreover, the mayor of Paris, Bertrand Delanoë, who was previously reluctant to the arrival of the Qataris, gave his approval and declared himself in favour of the acquisition of Paris Saint Germain by Qatar. The deal is concluded in June 2011 with the purchase by Qatar Sport Investment of the shares held by the majority US shareholder. Qatar Sport Investment becomes the majority shareholder of Paris Saint Germain. Its objective is to make Paris Saint Germain one of the five best clubs in the world. The objective is ambitious, especially as Paris

---

[58] http://www.footmercato.net/ligue1/psg-50-meur-de-dette-une-goutte-d-eau-dans-les-finances-de-colony-capital_46808, accessed 6 July 2013

Saint Germain does not enjoy a high reputation at European level. Before attracting truly renowned players such as Ibrahimovic, Paris Saint Germain is demonstrating its financial power on the transfer market. First of all with the purchase of Javier Pastore for 42 million euros, an Argentinian hopeful who did not win any major trophies at his former club in Palermo, Italy. Paris Saint Germain is betting on the future, making the biggest transfer in the history of Ligue 1 and showing the whole of European football that it can buy any player. Adding together the two transfer markets of summer 2011 and winter 2012, Paris Saint Germain is the most spendthrift club in Europe with cumulative spending in excess of €100 million. Qatar Sport Investment will become the owner of Paris Saint Germain after the purchase of the remaining 30% stake in March 2012. The President of Paris Saint Germain Al Khelaifi announces that Paris Saint Germain will invest €100 million a year in the transfer market. Leonardo, a former Paris Saint Germain player during the 1996-1997 season, has been the sports director symbolising the club's identity since the beginning of the project. Despite a second place at the end of the 2011-2012 season, Paris Saint Germain's ambition is confirmed with the purchase of internationally renowned players. The purchase of Thiago Silva from AC Milan for €45 million is a new record on the French transfer market. The purchase of Zlatan Ibrahimovic for 20 million euros has caused much controversy. Indeed, the player gets a salary reaching 14 million euros per year for three years. This is the highest salary in the history of the French Ligue 1. What isn't often mentioned in the press is that thanks to the 75% tax on income above one million euros per month, the French State receives 240 million over the three years of the contract at Paris Saint Germain[59]. Paris Saint Germain thus shows that it can attract the best players in the world at any price. So even when Monaco shattered the record for Ligue 1 transfers in 2013 by buying Colombian striker Falcao for 60 million euros, Paris Saint Germain replied by buying Uruguayan striker Edinson Cavani for 64 million euros. This shows that being the author of the biggest transfer in France is an important symbol for the Parisian club in its quest for world renown. Paris Saint Germain won Ligue 1 in 2013 and

---

[59] Nabil Ennasri, L'énigme du Qatar, Iris 2013, p.165

won the sum of 30 million euros[60]. In addition, participation in the Champions League each year brings in around €20 million for Paris Saint Germain.

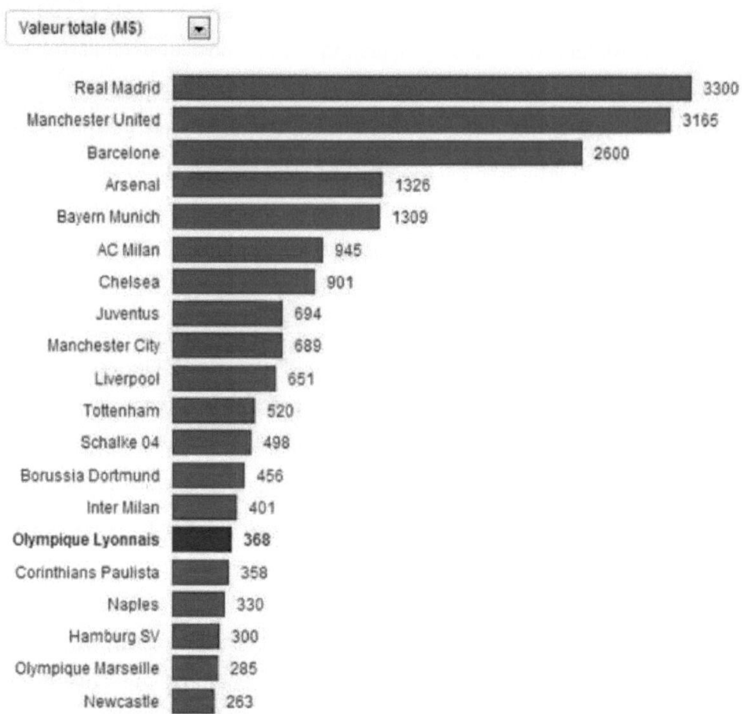

Figure 7: Ranking of the top 20 clubs according to their market value in 2012 extracted from the site :

http://www.leparisien.fr/sports/football/foot-le-real-madrid-devient-le-club-le- plus-chere-au-monde-18-04-2013-2737883.php

---

[60] http://www.ecofoot.fr/psg-250-millions-euros-recettes/, accessed 13 July

The Parisian club is faced with the pitfall of financial fair play, which forces clubs to respect a certain balance between income and expenses in order to take part in European competitions. Three solutions are available to clubs, including ticketing, sponsorship and derivative products. Paris Saint Germain's objective is to generate revenue of [61]€250 million from 2015. Since 2011, subscription prices have risen by 30% between the 2011-2012 and 2012-2013 seasons. The number of subscribers has also increased during this period.

period reaching 23,500 for the 2012-2013 season. For the 2013-2014 season, the number of subscribers is 31,600. Paris Saint Germain is still far from the best European clubs in terms of subscriptions. FC Barcelona has 86,000 subscribers for a 98,000-seat stadium, Real Madrid has 60,000 subscribers, and the club has a stadium with a capacity of 98,000. subscribers in an 86,000-seat stadium, Manchester United has 52,000 subscribers in a 75,000-seat stadium[62]. The problem could come from the size of the stadium which does not correspond to the standing that the Paris Saint Germain hopes to achieve. A solution under discussion would be the long-term acquisition of the Stade de France. This stadium is almost twice the size of the Parc des Princes, respectively 80,000 and 45,000 seats. It would double ticketing revenues and double the number of subscriptions. Another solution would be to enlarge the Parc des Princes. Qatar is in favour of expanding the stadium to 60,000 seats. This would require the Parc des Princes to be razed to the ground. However, Paris City Hall does not want the Parc des Princes to exceed 50,000 seats.

The second issue is sponsoring. While English clubs receive 150 million euros per year, French clubs only earn 50 million euros. In September 2012, Paris Saint Germain signed an image contract with the Qatar Tourism Authority running from 2012 to 2016. It will bring in €150 million a year for

---

[61] http://www.ecofoot.fr/psg-250-millions-euros-recettes/ , accessed 6 July 2013

[62] Http://www.lefigaro.fr/sport-business/2013/03/24/20006-20130324ARTFIG00073-le-qatar-impose-l-inflation-au-psg.php, accessed 6 July 2013

Paris Saint Germain. The problem raised is that the two entities have the same owner, the Qatari State. Moreover, the sponsor Fly Emirates could extend its contract with the Parisian club on condition that it increases the amount paid. It was 4 million euros per year and could reach 19 million euros per year until 2019. In addition, the equipment manufacturer of the Paris Saint Germain, Nike has a contract with Paris that expires next summer. Nike paid 6 million euros per year to Paris Saint Germain, which wants to renegotiate the contract upwards. A new competitor could take Nike's place. It is Burrda, a Qatari brand owned by Qatar Sport Investment. Burrda is already the equipment supplier of the promising Belgian national team. The Paris Saint Germain hopes to obtain a contract enabling it to obtain between 20 and 30 million euros per year. This is a sum that Nike is paying to Barcelona and Manchester United. Real Madrid receives 50 million a year from its equipment manufacturer.

Finally, the last issue is derivatives. Between the 2010-2011 season and the 2011-2012 season, sales of Paris Saint Germain rose from €17 million to €125 million. Then the price of the jersey increased by 13% between the 2011-2012 and 2012-2013 seasons, going from 75 euros to 85 euros. In 2012-2013, the arrival of David Beckham in winter 2013 for six months contributed to a 60% increase in jersey sales worldwide.

Qatar's investment in the Paris club is a long term project with the 2022 World Cup organised in Qatar in sight. Qatar is using Paris Saint Germain to gain legitimacy as the organiser of a World Cup in the face of an unconvinced European public. In addition, Qatar is developing a strategy to make considerable profits in the football industry with the development of the Burrda brand for example. These investments have nothing to do with the purchase of Malaga which is the fruit of the desires of the cousin of the Emir Abdullah bin Nasser Al-Thani who is not a member of the ruling family. He bought the club in 2010 for 32 million euros[63]. However, poor management of the club led Malaga to bankruptcy. The club could no longer

---

[63] Http://www.thenational.ae/sport/football/qatari-sheikh-buys-malaga-for-dh161m, accessed 8 August 2013

afford to pay the players' salaries and had to sell its best players at the end of last season. The club is 30 million euros in debt[64]. It was suspended from all European competitions by UEFA for four years in December 2012. In May 2013, the sanction imposed on Malaga was reduced to one year. The two clubs were managed in a totally opposite way.

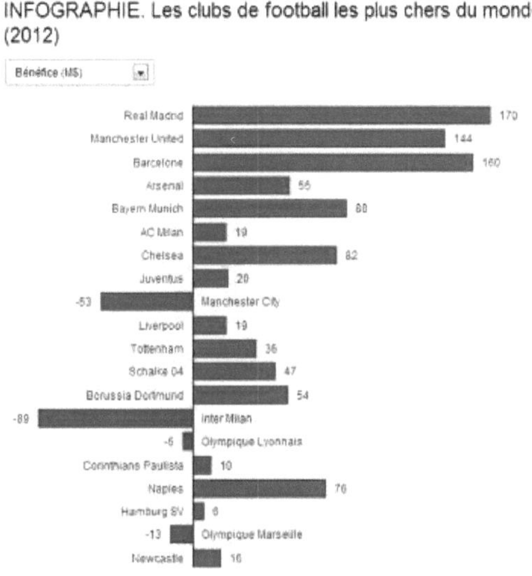

Figure 8: Profits of the top 20 clubs in terms of market value in 2012 extracted from the site

http://www.leparisien.fr/sports/football/foot-le-real-madrid-devient-le-club-le- plus-chere-au-monde-18-04-2013-2737883.php, accessed 8 July 2013

Finally, Qatar Sport Investment has entered the French football TV rights market. This is the second step after the acquisition of Paris Saint Germain. These investments are the result of a previous successful experience. Al Jazzera Sport, created in 2003, is becoming hegemonic in the broadcasting of sport in the Arab world. In 2009, Al Jazzera obtained the television rights

---

[64] http://www.metronews.fr/sport/malaga-bientot-rachete-par-tamim-bin-hamad-le-proprietaire-du-psg/mmcr!l0BbaTekM0gUc/, accessed 8 August 2013

to European competitions, major championships and the 2010 and 2014 World Cups for €650 million. This strategy is adapted to France. Al Jazzera Sport becomes BeIN Sport, a channel launched in June 2012. The president of Al Jazzera Sport- BeIN Sport is Nasser Al-Khelaifi, also president of Paris Saint Germain. He appoints Charles Bietry, former sports director of Canal Plus, as director of BeIN Sport. The idea is to transpose the television concepts of Canal Plus to this new channel. Numerous journalists are taken on from Canal Plus to give BeIN Sport a feeling of familiarity and continuity to viewers. The sports channel has obtained the television rights to broadcast 80% of League 1 matches, League 2 matches, 133 Champions League matches, half of the Italian and German championships and all of the Spanish and recently Portuguese championships and the 2012 and 2016 European Football Championships for 500 million euros[65]. The English championship is the only one to escape BeIN Sport. Canal Plus was forced to spend 200 million euros[66] to keep the English championship between 2014 and 2016. This championship exports its TV rights abroad for 6 billion euros[67]. Canal Plus denounces the unfair competition of BeIN Sport. Its subscription offer at 11 euros per month defies all competition whereas the subscription to Canal Plus is 40 euros per month[68]. In addition, BeIN Sport's revenues exceed 100 million euros while the cost of programming is around 300 million euros for the first year[69]. For the time being, BeIN Sport is running at a loss. The Qataris are not looking for short term profitability. Viewers have joined the BeIN Sport project. They were 1 million subscribers [70]six months after the launch of the channel.

---

[65] Nabil Ennasri, L'énigme du Qatar, Iris Editions 2013, p. 174.

[66] http://www.lefigaro.fr/medias/2013/02/01/20004-20130201ARTFIG00521-foot-anglais-canal-remporte-la-partie-face-a-bein-sport.php, accessed 8 July 2013

[67] ibidem
[68] ibidem
[69] http://www.france24.com/fr/20130711-canal-porte-plainte-contre-bein-sport-concurrence-deloyale-television, accessed 13 August 2013
[70] http://www.lexpress.fr/actualite/medias/deja-un-million-d-abonnes-pour-bein-sport-1-et-bein-sport-2_1184630.html, accessed 8 July 2013

Qatar's omnipresence on the sports media scene in France is in line with its desire to broaden its influence in order to make itself known. In sport, Qatar knows that it is not enough to obtain the organisation of the World Cup in 2022 to be recognised. To gain legitimacy, Qatar is establishing itself on the European continent, the epicentre of football and more particularly in France, which offers promising economic potential.

Qatar's investments in France should be put into perspective. Qatar is not seeking to buy France. The shares that it has obtained in large companies are for the most part only portfolio investments such as the 5% share that Qatar holds in Véolia. When it comes to direct investment abroad, Qatar's objective is not to become the owner of the company in question but to gain access to its board of directors. In this way, Qatar takes an influential position which enables it to conclude contracts to accelerate Qatar's internal development. Nor is France the European country where Qatar invests the most. The media coverage of Qatar's purchases in France distorts their true nature. The controversy over Qatar's ulterior motives only began in the early 2010's, culminating in the 2012 presidential campaign. The suburban fund was the subject of debate because it called into question one of the regalian functions of the state: to finance the development of its suburbs. France is losing momentum from an economic point of view because of the economic crisis and has been overtaken by the "Arab springs" that it failed to see coming. As for Qatar, its investments are not the whims of an ill-considered emir. It responds to a long-term strategy that aims to diversify the state's income and to secure this income. On the one hand, investments in France as well as in Europe are a means for Qatar to ensure its security. On the other hand, Qatar must develop a regional strategy that will enable it to emancipate itself from Saudi Arabia and become an important player in the geopolitics of the Middle East and the Maghreb to ensure its security. The growing mistrust in France would be better explained by the evolution of Qatar's foreign policy since 1996 in the Middle East and the Maghreb, an area of historical interest for France's foreign policy.

This second part tends to analyse the particularities of Qatar's foreign policy between 1996 and today, which respond to logics such as those of Richelieu's raison d'état, the French "great middle power" of the Cold War era or Kissinger's Realpolitik. Foreign policy will be limited to relations between Qatar and the states of the Maghreb, the Near and Middle East. The Malian crisis will also be evoked because it is the symbol of the divergence of interests between France and Qatar. From a diplomacy emancipated from its neutral, mediating and modern Saudi neighbour between 1996 and 2010 to an interventionist or even coercive regional policy from the "Arab Spring" beginning in 2011, the Qatari strategy responds to an imperative: to become the regional power of reference. Tools of influence such as the television channel Al Jazzera or mediation had enabled it to conquer Arab opinion by giving it an image of modernity and neutrality. Qatar's support to the Muslim Brotherhood following the "Arab Spring" led to a rejection of Qatar and its channel by Arab public opinion. This study aims to highlight Qatar's involvement in the main conflicts in the Middle East, the Maghreb and Mali recently.

Can Qatar take the leadership of the Sunni world?

A presentation of the geopolitical players in the Middle East and Maghreb is necessary to understand the context in which Qatar is extending its influence in these regions.

Until the Gulf War in 1990, Saudi Arabia was eligible for this role. They lost it when they demanded US intervention after Iraq's invasion of Kuwait. Several American military bases were set up in Saudi Arabia until 2003. Saudi Arabia has been facing internal political problems since 2010. The king's heirs are engaged in a succession struggle. Saudi Arabia is lagging behind on the regional scene. Iraq lost the war against Iran in 1988 and then the Gulf War in 1991. The fall of Saddam Hussein following the invasion led by the United States in 2003 left the country prey to internal conflicts between ethnic and religious fractions. Turkey is not recognised as a regional power because of its past as an Ottoman Empire and its desire to integrate the European Union. Egypt was the first Arab country to recognise Israel as a sovereign state through the Camp David Accords in 1978. Egypt was excluded for ten years from the Arab League and lost its influence in the Arab world. Following the "Arab Spring", Mubarak left power and relative internal political instability diminished Egypt's influence in the region. The decline of these regional powers benefits other countries in the Middle East, including Qatar.

Qatar's regional rivals are Saudi Arabia and Iran. Qatar has been an independent country since 1971. Border disputes with Saudi Arabia and Bahrain were only resolved in 2001. Qatar feels threatened and the example of Iraq's invasion of Kuwait in 1990 makes it fear for its security. In order to make its territory safe, Qatar is counting on the protection of the United States. In 2002, the United States set up a military base in El-Udeid. Qatar is the command centre for US military operations in the Middle East.

Qatar and Saudi Arabia are two Wahhabi kingdoms that compete to be the reference of the Sunni world. Saudi Arabia is the cultural reference of the Muslim world as it is home to the main holy places, including Mecca. Qatar is trying to obtain cultural legitimacy by using influence diplomacy, which will be analysed later. The conflicts in the Middle East will be influenced by this struggle between the historical reference and the emerging economic power for the leadership of the Sunni world. Qatar will support the Muslim Brotherhood while Saudi Arabia will support the Salafists. The Muslim Brotherhood are political organisations that embody for Qatar the future of political Islam. On the contrary, the Salafists, supported by Saudi Arabia, have shown their connivance with Al Qaeda since the attacks of 11 September 2001. Salafism has disqualified itself as a sustainable solution for the future. As soon as Qatar wishes to obtain the religious backing of Arab public opinion, Sheikh Al Qaradaoui appears to make inflammatory speeches called fatwas in favour of the choice of the Emir in foreign policy matters. He is a naturalised Qatari Egyptian who is one of the main artisans *"of the* Islamic legitimisation of the social and liberal reforms of the Emir and his wife", by acting as a counterweight in the country to the Salafists and to "a part of the Qatari Islamists, supporters of a literalist interpretation of Islam and hostile to any innovation (...)[71] ".

**Figure 9: Map of the Muslim world from
http://www.strategietotale.com/forum/17-l-expansion-arabe/87194-l-
expansion- de-l-islam-en-asia, accessed 30 August**

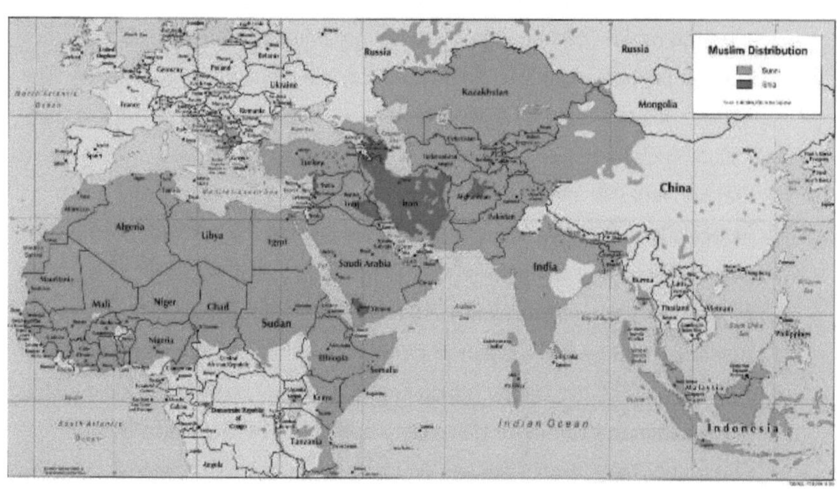

[71] Seniguer Haoues, "Qatar and the Islam of France: towards a new idyll? "Confluences Méditerranée, 2013/1 No. 84, pp. 101-115. DOI: 10.3917/come.084.0101

A- <u>Qatar's regional strategy: emancipation, influence and mediation</u>

1- Al Jazzera: the media as a tool for influence diplomacy

a- <u>The birth of Al Jazzera</u>

In 1995, the current Emir of Qatar Hamad Khalifa al-Thani seized power following a coup d'état. From 1996 the emergence of the Al-Jazzera chain embodies the will of Qatar's policy of influence. Finally, the "Arab Spring" which began at the end of 2010 marked a turning point for Qatar. It is an opportunity for Qatar to move from a policy of influence to an increasingly interventionist policy.

From 1996, Al Jazzera became the tool of Qatar's policy of influence. Al Jazzera fulfils two essential conditions for shaping Arab public opinion. It appears to be subservient to the authoritarian regimes of the Arab world and it aroused anti-American feelings in Arab public opinion during the wars in Afghanistan and Iraq in 2001 and 2003, which accentuated its legitimacy. Al Jazzera is the channel of the Arab people. It has shaped Arab public opinion around the Israeli-Palestinian conflict by defending the cause of the Palestinian people and showing their suffering.

b- <u>The emancipation of Al Jazzera, a regional benchmark</u>

c- <u>The internationalisation of Al Jazzera</u>

In 2006, the English version of Al Jazzera is launched. After becoming the most important media player in the Arab world, Al Jazzera competes directly with major Western news channels such as CNN and the BBC. An American satellite channel Middle East Television Network was launched in 2004 for the Arab world to compete with Al Jazzera. This channel is quickly a failure and this reinforces Al Jazzera as a symbol of freedom of expression for Arab public opinion. Al Jazzera played an important role after the attacks of 11 September 2001. It is the first channel to receive cassettes from Al Qaeda and the only one that can interview directly members of the terrorist organisation. During the American invasion of Afghanistan and then Iraq, Al Jazzera played the role of a major alternative media by showing images of the collateral damage to civilians caused by the American military operation. The premises of Al Jazzera in Baghdad are bombed by the Americans. Bush even threatened to destroy the premises of Al Jazzera in Doha. Al Jazzera thus proves that it is not a channel favourable to Westerners either. [72]

2- An independent foreign policy for Saudi Arabia based on neutrality and mediation

a- <u>Establish a balance of power against Saudi Arabia's neighbour by allying with Syria and Iran</u>

Qatar feels besieged by its powerful Saudi neighbour, which does not fail to recall its wishes for expansion on the territory of the Qatari emirate. In 1996, a few months after the coup d'état of Hamad Al Thani, Saudi Arabia is said to have financed a counter coup operation of a hundred men led by the Frenchman Barril. This operation was a failure. Qatar realises that it is necessary to establish a balance of power against the historic Sunni Egypt-Saudi Arabia axis. This is why Qatar is establishing diplomatic relations with Syria and Iran.

In December 2006, following the conflict in Lebanon, the Asian Games were organised in Doha. It was an opportunity for the three new allies to symbolically display their cooperation by sitting next to each other during the inauguration ceremony. Qatar is investing in Syria, around 6 billion dollars in the real estate and tourism sectors. An alliance with Syria allowed the development of a gas pipeline project linking Nabucco to the Gulf and Turkey. Qatar could have exported its gas directly to Europe without going

---

[72] ibidem

through the Strait of Hormuz and Iran. Bachar El Assad did not wish to offend his Iranian ally and did not follow up on this proposal by Qatar. Moreover, the port of Latakia is becoming a leading industrial platform thanks to Qatari investments. On the international scene, Qatar is militating for a return to favour of Bachar El Assad within the international community. France will take part in this campaign to rehabilitate Bashar El Assad by inviting him on 14 July 2008. Among the factors motivating this French support are the creation of the Union for the Mediterranean and the desire to distance Assad from his Iranian ally. In February 2010, a cooperation agreement was signed between Qatar and Syria[73] . A strong and almost family relationship is established between Qatar and Syria until 2011[74].

Geographically and economically, the proximity of Iran and Qatar is the imperative of their cooperation. The two countries share the operation of the North Dome. In 2004, an oil extraction platform in Qatar was destroyed by Revolutionary Guards. Since then, Qatar is the country in the Arab world with the best relations with Iran. Indeed, Qatar supports Iran's nuclear enrichment programme. Symbolically, Qatar was the only one of the 15 members of the Security Council to vote against a UN Security Council resolution in 2006. It gave Iran an ultimatum to suspend its uranium enrichment activities. [75]Moreover in 2008, the Qatari Prime Minister Ben Jassim declared during an official visit to Iran that "Qatar's position on the nuclear issue is in line with that of the Republic of Iran[76]". In 2009, following disputed presidential elections in the West, Qatar renewed its support for Mahmoud Ahmadinejad[77].

To exist on the international scene, Qatar must stand out from its Saudi

---

[73] Nabil Ennasri, L'énigme du Qatar, Iris Editions 2013, p. 81.

[74] C.Chesnot,G.Malbrunot, Qatar: Les secrets du coffre-fort, Michel Laffon, 2013, p.191-199 :

The Emir, Mozah and the Prime Minister are building a mini-palace on the road to Beirut and Sheikha Mozah is inviting the Syrian first lady to her daughter's wedding in 2010.

[75] http://www.iran-resist.org/article2438.html, accessed 16 July 2013

[76] Nabil Ennasri, L'énigme du Qatar, Iris Editions 2013, p. 81.

[77] ibidem

neighbour. To do this, Qatar is initially forming an alliance with Saudi Arabia's Shiite regional rivals: Iran and Syria. Moreover, these states do not have good diplomatic relations with the United States for different reasons. This allows Qatar to be the privileged mediator between Syria, Iran and the Western world. Thus Qatar is becoming a key player in the geopolitics of the Middle East, which reinforces its image of neutrality, modernity and mediator.

b- <u>Qatar and Israel: secular ties</u>

As in the case of its diplomatic relations with Iran, Qatar's approach towards Israel is at odds with the foreign policy of its Saudi neighbour. While Egypt had set a precedent with the Camp David agreements in 1978, Qatar went further. The rapprochement with Israel took place just after the coup d'état of Emir Hamad in 1995. In 1996, Qatar welcomed the first Israeli commercial representation in the Gulf. In 1999, the Emir of Qatar obtained Israel's permission to visit Gaza[78]. He is the first Arab Head of State to go there. The good relations with Israel enable Qatar to install a television antenna in Ramallah allowing Al Jazzera to cover the Israeli-Palestinian conflict in the media. However, as an Arab country, Qatar has distanced itself on several occasions. First during the second Intifada between 2000 and 2004 and then in 2009 following the Israeli "hardened lead" offensive. This "uncomplexed" relationship [79]with Israel is to the credit of Foreign Minister Hamad ben Jassim. In 2002, in the middle of the Intifada, he met Shimon Peres.

bypassing the mediation efforts of its Saudi neighbour who had proposed a peace plan. He proudly declared to Al Jazzera that "our relationship with Israel is official whereas most other states also have relations with Israel, but discreetly"[80]. This position allows him to be the defender of the oppressed Arab peoples in the Palestinian territories while maintaining

---

[78] Ibidem, p.85
[79] Ibidem, p.84

[80] Ibidem, p.86

cordial relations with Israel.

c- Is Qatar the mediator of the Middle East?

These privileged and exclusive relations with Iran, Syria and Israel allow Qatar to play a mediating role in regional conflicts. The examples of the crises in Lebanon in 2006 and 2008 and the Israeli-Palestinian conflict are the most telling examples.

At the end of August 2006, the Emir of Qatar was the first foreign head of state to visit Lebanon during the truce of the conflict between Israel and Hezbollah. Qatar took advantage of its good relations with Israel to become the first country to benefit from the breaking of the air blockade imposed by Israel[81] . Qatar is investing in the reconstruction of the conflict-ravaged territories of South Lebanon.

In 2008, Qatar put an end to a major political crisis in Lebanon. This crisis followed the cedar revolution that ended in 2005 with the withdrawal of Syrian troops from Lebanese territory. The assassination of Rafic Hariri, Prime Minister is the cause of the banishment of Bashar El Assad from the international community in 2005. For eighteen months, the pro-Western Lebanese party and Hezbollah were unable to reach an agreement to designate a president. Since 2006, an opposition camp has been set up in Beirut as a sign of protest. Opponents are demanding a revision of the electoral law and a government of national unity[82]. Various mediation attempts by the Arab League have failed. After five days of bitter negotiations in Doha, an agreement is reached with the Qatar[83] arbitration.

Mediation in the Palestinian territories of Qatar was only possible thanks to the good relations Qatar has with Israel. Israel is the only state authorised

---

[81] C.Chesnot,G.Malbrunot, Qatar: Les secrets du coffre-fort, Michel Laffon, 2013, p.172

[82] Http://www.rfi.fr/actufr/articles/101/article_66485.asp, accessed 23 July 2013
[83] Http://www.rfi.fr/actufr/articles/101/article_66473.asp, accessed 23 July 2013. This agreement redivides the constituencies. Beirut elects 19 deputies out of 128

to let the Qatari emir enter these territories during his surprise visit in October 2012. Hamas has ruled over the territories of the Gaza Strip since 2007. It does not recognise the State of Israel and it is not recognised by Israel or by Western countries. Hamas finds itself in a situation of isolation. It is Emir Hamad who breaks this situation of isolation on 23rd October 2012. He is the first head of state to meet officially with Hamas. The head of Hamas Khaled Mechaal is a friend of the Emir who resided in Doha[84]. The Emir is going to Gaza with the promise to invest 400 million euros for road improvement projects and agricultural development[85] . Moreover Qatar is working to influence the post-Abbas period and make Mechaal his successor. Qatar is bringing Hamas back into the fold of the Sunni axis. The financing of Hamas becomes mainly the responsibility of Qatar. Hamas moves away from Iranian influence. Qatar succeeds in bringing Hamas back into the Sunni fold. It becomes its main creditor in front of Iran[86].

Thanks to its special relations with Syria, Iran and Israel, Qatar is implementing a policy of mediation which makes it a leading player in the Middle East. Qatar becomes the mediator that puts an end to important regional crises such as those in Lebanon. However, as the latest example of the Gaza Strip shows, Qatar favours Islamist political parties. This is a trend that has its origins in the beginning of the "Arab Springs". The "Arab Springs" are an opportunity for Qatar to take leadership in the Arab world. All the alliances mentioned above are called into question in the name of the old rivalry between Sunnis and Shiites. And even within Sunnism, the struggles for influence are intense between Saudi Arabia, a historical reference, and Qatar, an emerging regional power. Has Qatar encouraged a movement

of emancipation of the peoples of the Maghreb and the Middle East or did

---

[84] Http://www.lefigaro.fr/international/2012/10/23/01003-20121023ARTFIG00323-l-emir-du-qatar-affiche-son-parti-pris-pro-hamas-a-gaza.php, consulté le 5 août 2013

[85] Http://www.rfi.fr/moyen-orient/20121023-gaza-emir-qatar-cheikh-hamad-ben-khalifa-al-thani-hamas-haniyeh, accessed 5 August 2013

[86] http://www.lemonde.fr/proche-orient/article/2012/10/24/l-emir-du-qatar-brise-l-isolement-diplomatique-du- hamas_1780041_3218.html, accessed August 5

it deliberately lead them towards political Islamism?

B- <u>Qatar's influence on the "Arab Spring" and the beginnings of its interventionist policy</u>

1- Is Qatar the defender of the freedom of the Arab peoples?

a- <u>Tunisia and Egypt: Qatar's policy of influence at its peak</u>

Qatar, through its tool of influence Al Jazzera, acted in the shadows during the first Arab revolutions in Tunisia and Egypt.

Al Jazzera is the channel of the Arab people and Qatar sided with the demonstrators very early on. The live broadcasting of demonstrations and repression in Egypt and Tunisia gave rise to a feeling of injustice and revolt which amplified the movement. The success of these growing demonstrations led Western leaders to disavow the authoritarian regimes in Egypt and Tunisia. In a few days, Mubarak and Ben Ali went from President to dictator in the speeches of Western heads of state. Mubarak and Ben Ali leave power. These regimes were perceived by opponents and public opinion as corrupt regimes, under the protection of the United States and Western states. The elections held in Egypt and Tunisia are in line with the desire of the United States for a democratic transition. The Islamist parties of the Muslim Brotherhood in Egypt and Ennahda in Tunisia won the elections thanks to support from Qatar. It is at this point that Qatar loses its attribute of regional mediator by taking a stand for the Islamist political parties and among them the Muslim Brotherhood. Of course these parties came to power through the elections. However, Qatar, by financing these parties, put them in a position of strength.

They were the best structured parties to come to power as opposed to the extremely divided opposition parties.

In Egypt, Qatar's support to Mohammed Morsi's regime has resulted in investments of six billion euros[87]. Qatar is the biggest contributor in Egypt among the Gulf States[88]. However, the recent fall of Morsi on 3 July 2013 is a setback for Egyptian diplomacy which loses an allied regime. This setback took Qatar by surprise and it reacted timidly through the voice of the spokesman of the Ministry of Foreign Affairs assuring that Qatar "will continue to support Egypt. We respect the will and choices of the Egyptian people. [89]

According to the Syrian Foreign Minister, Qatar would have funded the Tunisian Ennahda party following Gannouchi's visit in 2011. According to this source, the amount of funding would be 150 million pounds[90]. These allegations have been denied by the leader of the Ennahdha party. Although there is no proof of the amount of funding, there is no doubt about Qatar's financial support to Ennahdha. Al Jazzera's media support to the Tunisian and Egyptian governments among others has made it lose its neutrality. This channel, which wanted to be the voice of the people, has fallen into political-religious propaganda so much so that the Egyptian and Tunisian populations distrust it. Al Jazzera now suffers from the same symptoms as the national channels before: lack of independence with the appointment in 2011 of the Emir's cousin, Hamad Bin Thamer Al Thani to the presidency of the board of directors of Al Jazzera, taking political-religious positions very marked in favour of political Islamism. Al Jazzera is no longer the channel of the people.

In Yemen, Qatar is urging President Saleh to step down in 2011. Qatar has not encouraged all revolt movements in the Arab world.

---

[87] http://www.courrierinternational.com/article/2013/07/11/revers-pour-le-qatar, accessed August 25, 2013
[88] ibidem
[89] http://www.rfi.fr/moyen-orient/20130705-egypte-reaction-mesuree-qatar-chute-mohamed-morsi-al-thani, accessed 25 August 2013
[90] http://www.kalima-tunisie.info/fr/News-Ennahda-le-Qatar-n-a-pas-finance-notre-campagne-item-3227.html, accessed 17 August

b- <u>The counter-revolution in Bahrain</u>

In Bahrain, a member of the Gulf Cooperation Council, events took place between February and March 2011. The Bahraini Shiite demonstrators, who represent 70% of the population, demanded reforms as in Tunisia and Egypt. They are repressed by the Sunni regime in Bahrain. Qatar and Saudi Arabia support the Bahraini regime for fear of the spread of revolts on their territories. Al Jazzera does not show any image of the repression in Bahrain and this event is passed over in silence. Qatar and Saudi Arabia agree to send troops to protect the monarchy of Bahrain. Saudi troops are still in Bahrain to protect the ruling regime. Saudi Arabia and Qatar are acting together to stem the Shiite threat. [91]

Qatar uses soft power to successfully carry out a policy of influence based on diplomacy, its economic power and an indirect cultural tool, Al Jazzera, while taking care to censor the internal opposition. However, the backlash of these positions following the "Arab Spring" is being felt within Arab public opinion. Indeed, the Qatari strategy is evolving during the crisis in Libya. Qatar is no longer a neutral player and mediator in the region. It seeks to put in power allied regimes that serve its interests. Thus the crisis in Libya allows Qatar to move to the second stage of its evolution. It is abandoning its policy of influence for a more interventionist policy to the detriment of its popularity among Arab public opinion and its image of neutrality and modernity which it is trying to convey in Europe and which it has conveyed for more than a decade in the Arab world.

---

[91] http://www.nybooks.com/articles/archives/2011/oct/27/strange-power-qatar/?pagination=false, accessed 17 August 2013

2- Libya and Mali: Qatari interventionism and the antagonism of Franco-Qatari interests

a- <u>Libya: Qatar's new interventionist strategy</u>

After the success of the "Arab Springs" and its policy of influence in Egypt and Tunisia, Qatar adopted an increasingly interventionist policy in Libya. It was Sheikh Al Qaradaoui who managed to convince the Emir to choose the side of the insurgents.

In Libya, Qatar supports insurgents against Gaddafi's regime. He is an emblematic figure in the Arab world. He is the defender of a certain vision of pan-Arabism. Qatar is the second state after France to recognise the legitimacy of the Libyan National Transitional Council (CNT). This support is no longer just diplomatic. For their intervention in Libya to be legitimate, NATO needs the military support of Qatar and the Arab League. Arab countries must be able to intervene militarily in an Arab country in a state of civil war.

Qatar participates in the NATO-led military operation in Libya. It provides air and land forces to NATO. Six Qatar Mirage 2000[92] aircraft are made available to NATO; this is half of the Mirage 2000 Qatari fleet. France delivers anti-tank missiles to Qatar. These are 8,000 surface-to-air miss[93]iles that disappear from the Libyan army's stocks following the defeat of Gaddafi. The Qatari ground military forces sent to Libya are said to have numbered 5,[94]000. This questionable figure must include special forces, soldiers and mercenaries. The CNT has received 5 out of 18 air convoys of arms from Qatar. While the CNT was supposed to receive help from Qatar, Qatar has armed carefully selected Islamist groups. Among them, the group

---

[92] C.Chesnot,G.Malbrunot, Qatar: Les secrets du coffre-fort, Michel Laffon, 2013, p.184

[93] ibidem

[94] Http://blog.lefigaro.fr/malbrunot/2011/11/5-000-forces-speciales-du-qata.html, accessed 24 August 2013 :

This figure should be qualified because Mr Malbrunot's source speaks of troops made up of hundreds of soldiers throughout Libya.

of Ali Al Salabi, a friend of the Qaradawi Sheikh of the Muslim Brotherhood, the group of Behladj and a group accused of murdering a general in July 2011 called the Katiba Obaida Ibn Jarrah [95]. The lack of coordination within NATO between the West and Qatar in particular is beyond doubt. Qatar has been reprimanded by the French President over this very controversial aid[96] which feeds the idea that Qatar has a hidden agenda.

After the fall of Gaddafi's regime, Qatar finances and supports more particularly Islamist opponents who are members of the Libyan National Transitional Council. The objective is similar to that already achieved in Tunisia and Egypt: to promote the election of the Muslim Brotherhood in the 2012 legislative elections. Personalities such as the former jihadist Behladj are trying unsuccessfully to convert to politics[97]. It is finally the most liberal party led by Mahmoud Djibril, former chief executive of the CNT[98], which won the elections to the great displeasure of the Muslim Brotherhood, despite the support of Qatar.

Western and Qatari support for the Libyan rebels led indirectly to the conflict in Mali. This conflict is of interest because it stems from the failure of disarmament in Libya led by the CNT. Armed groups continued their struggle in the Sahel. The Malian crisis is an opportunity to show why the two economic partners, France and Qatar, have conflicting interests in Mali.

b- <u>The Mali crisis: an indirect confrontation between France and Qatar?</u>

Qatar is accused of financing and arming terrorist organisations in northern Mali. These are the Mujao and Al Qaeda in the Islamic Maghreb (AQIM). While there is no material proof of these financings, the presence of Qatar in Mali is proven. The humanitarian operations carried out by Qatar arouse

---

[95] http://blogs.mediapart.fr/blog/arnaud-castaignet/071211/le-qatar-un-inquietant-electron-libre-en-libye,consulté August 24, 2013

[96] C.Chesnot,G.Malbrunot, Qatar: Les secrets du coffre-fort, Michel Laffon, 2013, pp.188-189

[97] Http://www.lefigaro.fr/international/2012/07/18/01003-20120718ARTFIG00496-libye-deroute-electorale-du-parti-des-freres-musulmans.php, accessed 25 August 2013

[98] Http://www.lefigaro.fr/international/2012/07/18/01003-20120718ARTFIG00489-mahmoud-jibril-s-impose-a-un-pays-qui-veut-oublier-kadhafi.php, accessed 25 August 2013

suspicion. In July 2012, the Qatari Red Crescent was one of the only NGOs to have intervened in Gao, then controlled by the Mujao[99] . A financial aid of 3 billion CFA francs is promised to the inhabitants of the city as well as food aid[100]. The inhabitants do not see any financial aid promised. The working methods of Qatari NGOs are contested. They work in isolation without cooperating with Malian NGOs. They are suspected of using these humanitarian operations as a cover to finance terrorist groups such as the Mujao and Ansar Dine. This

financing whether it is deliberate or the result of a misappropriation of aid is confirmed by the Canard Enchainé of 6 June 2012.

Qatar has not declared itself in favour of the "Serval" operation led by France in January 2013. Just before the start of operations, two Qatari aircraft reportedly dropped weapons to the rebels in North Mali. There are many issues at stake which would push Qatar to act in the shadows to help terrorist groups. On the strength of its victory in Libya, Qatar intends to widen its zone of influence in the Sahel region. Mali is thus a major stake. Qatar can become an intermediary between an Islamic state in northern Mali, the Malian government and France. Moreover, Mali has gas reserves that require infrastructures that Qatar already has and its subsoil is rich in oil and uranium[101] . Obviously, Qatar denies these allegations outright. It is absolutely not in Qatar's interest to turn the Mali crisis into an open conflict with its French ally.

The interventionist stage is a transition to a coercive policy from Qatar to Syria. If previous conflicts had at stake the leadership of the Sunni world in the Middle East and the Maghreb, the crisis in Syria is redrawing the divide

---

[99] Http://www.rfi.fr/afrique/20121102-presence-qatar-nord-mali-doutes-persistent-mujao-mnla-dioncounda-traore-niger-aide-humanitaire-cooperation, accessed 28 August 2013
[100] ibidem

[101] Http://www.lexpress.fr/actualite/monde/afrique/le-qatar-intervient-il-au-nord-mali_1194852.html, accessed 28 August 2013

between Sunnis and Shiites.

C- <u>The Syrian conflict: the clash between Qatari interests and Western and French interests more particularly</u>

1- The return of the regional cleavage between Sunnis and Shiites

During previous "Arab Spring" events, Qatar sought to provide leadership to Sunni countries. It is another logic that animated Qatar during the Syrian conflict. On a global scale, it is a

opposition between the United States, France and the United Kingdom on the one hand and Russia and China on the other. At the regional level, the cleavage between Sunnis (80% of Muslims) and Shiites (20% of Muslims) is taking shape again. The Shiite axis is composed of Iran, Syria and Hezbollah in Lebanon. The Sunni axis is led by Qatar. Egypt is in a period of political instability following the fall of Mubarak in 2011 then Morsi who was deposed in 2013 by the army. The other historic Sunni power, Saudi Arabia, is in the middle of an internal war of succession. The crisis in Bahrain showed that Saudi Arabia and Qatar were ready to ally despite their rivalry against the Shiite threat. The Syrian conflict confirms this trend. In Syria the Alawites, the heretical branch of Shiism according to the Sunnis, represent a minority of 10%, i.e. two million inhabitants. Bachar El Assad has been in power since 2000. He is Alaouite and relies on an alliance between the Christian and Alaouite minorities to consolidate his power against the Sunni Muslims who represent about 80% of the Syrian population. It would be very advantageous to topple Bashar El Assad in order to establish a Sunni regime in Syria. A Sunni regime would join the Sunni alliance led by Qatar against Iran. Syria remains Iran's sole Arab ally. If it is led by a Sunni majority then Iran would find itself isolated without an ally in the Arab world.

2- Qatar, "capital of the opposition" to the regime of Bashar El Assad

a- The Syrian National Coalition: the weapon of Qatar's diplomacy of influence in Syria?

In spring 2011, Qatar is trying to play its role as a mediator to urge Bashar El Assad to stop the bloody repression[102]. The Syrian Foreign Minister met Emir Hamad who promised to work for reconciliation in his country through Al Jazzera and to invest billions of dollars in the renovation of his country[103]. This attempt at mediation was a failure and six months after the start of the revolts in Syria, Qatar put an end to its alliance with Syria. The Syrian embassy in Qatar is closed. Once again encouraged by

Sheikh Al Qaradaoui to take action against Bashar El Assad, Qatar is in the front line of an Arab diplomatic force, the Arab League to urge the Syrian head of state to leave power. It is also following the advice of its adviser Azmi Bishara, head of the think tank Arab Center for Research and Policy Studies[104]. Qatar used its influence to obtain Syria's exclusion from the Arab League[105]. In November 2012, the opponents of Bashar El-Assad met for the first time in Doha and decided to form a coalition with a single representative, the Syrian National Coalition (SNC). This representative, Al-Khatib, visited Qatar in 2003 and 2004. He resigned in May 2013[106]. It is a diplomatic success for Qatar which proves that it is an influential regional power despite the size of its territory and its weak military power.

In the NSC, Qatar once again supports the Syrian Muslim Brotherhood led by Farouk Tayfour one of the three vice-presidents of the NSC. In July 2013 elections are held to appoint the new representative of the NSC. The Qatari-

---

[102] C.Chesnot, G.Malbrunot, Qatar: Les secrets du coffre-fort, Michel Laffon, 2013, p.197

[103] C.Chesnot, G.Malbrunot, Qatar: Les secrets du coffre-fort, Michel Laffon, 2013, pp.197-198

[104] http://www.jeuneafrique.com/Article/JA2743p015.xml0/, accessed 2 September 2013

[105] Ibidem, p 200

[106] Http://www.lefigaro.fr/international/2013/03/25/01003-20130325ARTFIG00433-l-opposition-syrienne-devra-se-restructurer.php, accessed 28 August 2013. Al Khatib was against a partition of the country and tried to reach out to Bashar El Assad

backed candidate, Mustapha Sabbagh loses the vote by 3 votes to a consensus candidate, Ahmad Jarba. The project of a transitional government led by Qatar's support, Hitto, did not see the light of day[107]. Indeed the Free Syrian Army (ASL) had bad relations with the CNS. Composed of the Al Nosra group and Islamist and Salafist rebels, the ASL does not recognise the legitimacy of the CNS as the representative of the Syrian people.

b- <u>Syrian conflict: the struggle for influence among Sunnis</u>

At the same time, Qatar is selling arms to the Syrian rebels and is supporting them logistically from summer 2012 after the failure of the UN's Annan mission. France is responding favourably to Qatar's request to provide logistical support to the Syrian rebels with around 14 million euros of expenditure including night vision equipment and radios[108]. This money paid to the CNS is not intended to enable them to buy arms. Qatar is said to have spent \$3 billion [109]to arm the Syrian rebels. The free Syrian army is very heterogeneous. The jihadist group Al-Nosra, a Syrian offshoot of Al Qaeda in Syria has joined the Syrian rebels on the ground[110] . This armed group was placed on the list of terrorist organisations by the United States in December 2012. On the ground, we find the struggle for influence between Saudi and Qatari neighbours. Some support the Salafists and others the Muslim Brotherhood.

Moreover, on a global scale, Qatar acts with the diplomatic support of the United States and Western countries such as France. This cooperation is flawed because Qatar's objectives are not shared by its allies. The United States has another agenda that would allow the Baathist regime to survive

---

[107] Http://www.courrierinternational.com/article/2013/04/18/le-qatar-nous-impose-le-premier-ministre, accessed 28 August 2013. Hitto who lives in Texas becomes prime minister of the short-lived transitional government against the advice of the NSC.

[108] C.Chesnot, G.Malbrunot, Qatar: Les secrets du coffre-fort, Michel Laffon, 2013, p. 206
[109] http://www.latribune.fr/actualites/economie/international/20130517trib000765147/syrie-le-qatar-aurait-depense-3-milliards-de-dollars-pour-armer-les-rebelles.html, accessed 28 August 2013

[110] C.Chesnot, G.Malbrunot, Qatar: Les secrets du coffre-fort, Michel Laffon, 2013, p. 206

the civil war if Bashar El Assad agreed to resign. Thus, the United States would maintain in power a bassist regime that has never attacked its ally Israel despite the animosity of their relations. This would be less risky than encouraging the emergence of a Sunni regime hostile to Israel and the West.

Syria is responding to this Sunni axis. At the regional level, Syria has tried to break the Sunni alliance between Saudi Arabia and Qatar by using its electronic army to hack into the Al-Arabiya television site in April 2012. Information circulated there about a hypothetical coup d'état targeting the Emir of Qatar. In addition, Iraq is said to have evidence that Qatar is providing military and financial aid to Al Nosra[111] . These rumours are proving compromising for French leaders who find themselves fighting terrorists in Mali and indirectly helping them in Syria because they are allied with Syrian rebels to bring down the regime of Bashar El-Assad.

Qatar is fully engaged in the civil war in Syria. Between a policy of influence to help set up the CNS, behind-the-scenes action to help the Syrian rebels and proven military involvement with the installation of the CNS, Qatar has been a major player in the Syrian civil war.

of training camps for young fighters reinforcing the Syrian armed rebellion, Qatar has really taken the leadership of the Sunni axis in the struggle to bring down the regime of Bashar El Assad.

---

[111] ibidem p. 208

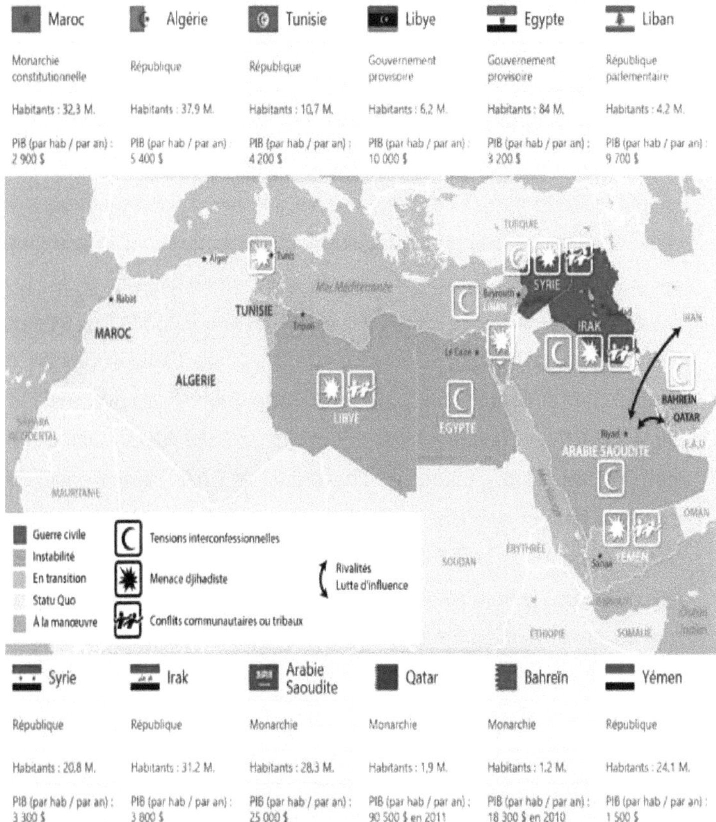

Figure 10: Provisional assessment of Arab Spring in September 2013
http://www.lexpress.fr/actualite/monde/proche-orient/infographie-que-reste-t-il- des-printemps-arabes_1277890.html, accessed September 5

Qatar's geopolitical strategy is ambiguous. It uses Al Jazzera to promote freedom of expression in the Arab world outside its borders while censoring the Qatari opposition. It supports Arab revolutions in Yemen, Tunisia and Egypt officially in the name of freedom but helps to suppress the revolt of the people of Bahrain. It is the ally of the United States, an intermediary between the United States and the Arab world, but it supports and finances Islamic parties that are not very favourable to the Americans. During the

"Arab Springs". Prince Al-Thani sought to get rid of emblematic and influential personalities of the Arab world such as Mubarak and Gaddafi to replace them.

During the Arab springs, two phases are distinguished. The first phase are the peaceful revolutions in Tunisia and Egypt. Qatar used its power of influence, diplomacy, economy, culture and the dictators withdrew without civil war. The second phase concerns the violent revolutions in Libya and Syria where Qatar intervened militarily in a climate of civil war. In Libya and Syria, Qatar appears as a limited coercive power. It needs to cooperate with international organisations in order to succeed. Qatar's support for the Muslim Brotherhood in the Arab world made them accomplices of the "Arab winter". In Egypt, following the fall of Mohammed Morsi, Qatar is the big loser of the military takeover. The arrival of Emir Tamim could call into question certain aspects of Qatar's regional strategy. Indeed the Prime Minister and Minister of Foreign Affairs Hamad Ben Jassim was dismissed from his post when he was the emblematic figure of Sheikh Hamad's regime. He embodied the strategy of mediation notably in his unashamedly open relations with Israel. This transfer of power, unprecedented for a Gulf country, allowed the arrival of new leaders who were more moderate and less interventionist[112]. The new Prime Minister is Abdellah Ben Nasser Al Thani and the new Minister of Foreign Affairs is Ahmed Ben Jassim, former director of the channel Al Jazzera. By assigning these two posts to different people, the young Emir puts an end to the concentration of power from which Hamad Bin Jassim benefited. Will these changes lead to more moderation in the regional policy of a Qatar that is beginning to pay the price of its support to the Muslim Brotherhood in the Middle East and Maghreb countries?

---

[112] http://www.rfi.fr/moyen-orient/20130626-qatar-nouvel-emir-remercie-premier-ministre-ancien-bras-droit- pere-tamim-ben-hamad-al-thani consulted on September 2

Conclusion

Qatar has experienced a meteoric rise in power over the last twenty years. The exploitation of natural gas has enabled Qatar to develop its economy. Qatar has become a fragile rent-based economy because it is too dependent on gas exports. The objective of the Qatar National Vision for 2030 is to diversify the income of the Qatari state so that it can become a developed state. Among the pillars of this strategy, the analysis of Qatari investments in Europe and more particularly in France has enabled us to understand that Qatar's strategy is based on the long term. It is based on long-standing diplomatic and personal ties and a love of French culture in Qatar. The media coverage of Qatar's investments in France is disproportionate to the reality of these investments, which must be put into perspective. France is far from being the only European State in which Qatar invests. Nor is France the State in Europe where Qatar invests the most. Qatar's investments are carefully thought out. It is necessary to invest in areas of excellence such as industry, real estate and sport in France. Opportunities to make profitable investments and long-term profits are facilitated by the period of economic crisis that has hit Europe since 2008 and by the tax advantages granted to Qataris. Again, these tax advantages already existed for other emirates and for Kuwaitis in the 1980s. Qatar can thus secure its oil windfall and diversify the State's revenues in order to develop. These investments also enable Qatar to conclude defence agreements with Western states to secure its territory. Qatar wishes to extend its influence in a region where France has long had a great deal of influence. This could explain why there is more concern in France about Qatar.

The securing of the territory is the leitmotiv which has animated the investment strategy of Qatar and its regional foreign policy since 1995. This foreign policy is characterised by the will of Qatar to be an inescapable regional power. This explains why Qatar has allied itself with Syria and Iran.

Qatar acts as an intermediary between the major Western powers, France and the United States, and those countries, Iran and Syria, which had poor relations with each other. These exclusive relations, such as those Qatar has with Israel, allow it to intervene in the conflicts in the Middle East as a mediator. The "Arab Springs" are the turning point of this promising and daring foreign policy. Qatar cultivated an image of modernity because it defends the freedom of political opinions through Al Jazzera and neutrality through its reputation as a mediator. With the "Arab Springs", Qatar supports movements for the emancipation of peoples towards democracy. Qatar uses financial means so that the political Islamism embodied by the Muslim Brotherhood are the best structured parties to access power. Qatar also uses Al Jazzera to make propaganda in favour of these political parties. Already Qatar and Al Jazzera are arousing mistrust among the peoples of Tunisia and Egypt for example. Qatar supported the Arab winter which brought Ennahda and the Muslim brothers to power. The fall of Morsi in Egypt is the symbol of the Egyptians' rejection of political Islam. It is a setback for Qatar. Moreover, Qatar has stubbornly pursued this path by its interventionism in Libya in order to favour the election of Muslim brothers who finally failed to come to power. The intervention in Libya has had negative effects in the Sahel region and particularly in the north of Mali. Terrorist organisations were able to take advantage of weapons that had disappeared from Libya. It is the Malian conflict that shows most clearly the divergence of interests between France and Qatar in the region. Qatar was suspected of financing terrorist organisations such as Mujao relatively unintentionally. The Syrian crisis is the conflict that awakens all the divisions. The Cold War divide between the United States and Russia on a global scale. On a regional scale, the Sunni-Shiite divide has been the dividing line in the Syrian conflict. On a local scale, it is a civil war. The "Arab Springs" swept away the emblematic figures of Arab nationalism, Ben Ali, Mubarak and Gaddafi. Paradoxically, Qatar finds itself in a position of impotent leadership of the Sunni world. Impotent because the aid provided to the Syrian rebels is insufficient to defeat the regime of Bashar El Assad. Only a UN or NATO intervention could reverse the balance of power in favour of the Syrian rebels. The United States and France have recently shown their determination to intervene without UN Security Council agreements and without their traditional British allies. Qatar's regional foreign policy has become the political backing for its investments in Europe, as if to convey the message that Qatar has fully embraced globalisation and international trade without forgetting its culture and religion. The Qatari leadership in the Sunni world is relativised by the cultural vacuum it represents: it is not yet a cultural and religious reference as can be Saudi Arabia with Mecca. What will happen to Qatar when Egypt and Saudi Arabia, one in a period of political instability, the other in a period of succession struggle, regain their roles as regional powers?

Bibliography

Books

Lazar Medhi, Qatar today, Michalon, 2013

Chesnot Christian, Malbrunot Georges, Qatar: The secrets of the safe, Michel Laffon, 2013

Ennasri Nabil, L'Enigme du Qatar, éditions Iris 2013 Boniface Pascal dir,

L'Année stratégique 2013

Verdez Gilles, Hermant Arnaud, Le PSG, Qatar and money: the forbidden investigation

Magazines

International Issues, Printemps arabes, n°53, January-February 2012, pp

4-86 Challenges, n°350, 20-26 June, pp 52-62

Sitography

MichelLazar , a policy of influence :
http://www.diploweb.com/Qatar-une- politique-d-influence.html

Aissaoui, LeQatarpetitémiratetgrandesambitions :
http://nadia- aissaoui.blogspot.fr/2012/02/le-qatar-petit-emirat-et-grandes.html

Eakin Hugh, Strange power of Qatar :
http://www.nybooks.com/articles/archives/2011/oct/27/strange-power-qatar/?pagination=false

HealiryCecily, What' sdrivingqatar' sforeignpolicy :
http://www.voanews.com/content/whats-driving-qatars-foreign-policy/1524917.html

TheriseofQatarPigmywiththepunchofagiant :
http://www.economist.com/node/21536659

Rocco, Anne-Marie , But where do you think the billions of Qatari dollars go ? :
http://www.challenges.fr/monde/20120216.CHA3244/mais-ou-vont-les-milliards-du- qatar.html

Drief Zineb, interview with Mr. Bitar, Qatar: "A poor Arab is an Arab, a rich Arab is a rich".
http://www.rue89.com/rue89-eco/2012/03/14/qatar-un-arabe-pauvre-est-un-arabe-un- arabe-riche-est-un-riche-229641
http://www.diplomatie.gouv.fr/fr/dossiers-pays/qatar/presentation-du-qatar/article/composition-of-government-2036
http://www.lenouveleconomiste.fr/financial-times/cheikh-al-thani-ancien-premier- minister-of-qatar-all-you-call-it-not-true-30657/

Press

Le Monde Le Figaro

The Diplomatic World

Table of Contents

Annexes

Interview with Mr Nabil Ennasri, 25 July 2013

1-   **How does the transition from a mediator role to a more coercive (Libya and Syria) and committed (Egypt and Tunisia) strategy affect Qatar's image of modernity and neutrality**?

I don't think that Qatar's image is fundamentally affected by Qatar's activism as the emirate still keeps the issue of mediation among its main diplomatic activities. As we can see in Darfur, Qatari diplomacy has not changed and even with countries such as Ethiopia with which relations were tense, a diplomatic approach of conciliation has enabled the establishment of strong diplomatic ties (see here in particular

:

http://studies.aljazeera.net/en/reports/2013/05/2013528104610736526.ht m). Now, what is certain is that Qatar is still blaming the blow of a voluntarist diplomacy which, in my opinion, cannot be summed up as support for the Islamists in the strict sense of the word. Qatar has given its support to processes of democratic transition that have brought Islamist formations to power, and it is this sequence that, in my view, must be understood.

2-   **Is it this new strategy that is the main source of fear in France (investments, patronage, aborted Suburban Plan)? Is there a detailed document of this aborted suburban plan?**

The fear of Qatari investments in France is due to a combination of factors. France has a problem with money and a problem with Islam. It is these two problems that have found their crystallisation with Qatar. In addition, there is a diffuse feeling of being downgraded as a result of globalisation and it is also this fear of the future and of the change in the world's course which is expressed through the feeling towards Qatar. Finally, there is also the dazzling ascent which makes you feel uncomfortable. Few countries have been able to emerge in such a way and in such a short time and this exceptional trajectory inevitably provokes jealousy and tension in some of them   . No, there was no detailed plan of the aborted suburban plan.

3-   What are the risks/opportunities in France in concrete terms? Is France (5 million Muslims) becoming the scene of a cultural rivalry between Saudi Arabia and Qatar for Sunni leadership?

This may be the case in the Middle East but not in France. Both countries know that they are walking on eggshells when it comes to issues related to Muslims in France. They have no interest in carrying out charming offensives for populations who, moreover, wish to leave the supervision of the consulates to achieve real independence.

4-   Can the Qataris get tired of France's lack of recognition and leave?

I don't think so, even if it is a risk that has been mentioned here and there. Relations between the two countries are the basis of a strategic partnership and therefore do not depend fundamentally on the image of this or that country. It is true that the bad press in Qatar is damaging, but not at such a level that it would call into question the fundamentals of a relationship based on constantly increasing military, political and cultural exchanges.

I want morebooks!

Buy your books fast and straightforward online - at one of world's fastest growing online book stores! Environmentally sound due to Print-on-Demand technologies.

Buy your books online at
www.morebooks.shop

Kaufen Sie Ihre Bücher schnell und unkompliziert online – auf einer der am schnellsten wachsenden Buchhandelsplattformen weltweit! Dank Print-On-Demand umwelt- und ressourcenschonend produziert.

Bücher schneller online kaufen
www.morebooks.shop

KS OmniScriptum Publishing
Brivibas gatve 197
LV-1039 Riga, Latvia
Telefax: +371 686 204 55

info@omniscriptum.com
www.omniscriptum.com

Printed by Books on Demand GmbH, Norderstedt / Germany